Environmental Humanities and Theologies

Many ways of thinking about and living with 'the environment' have their roots in the Bible and the Christian cultural tradition. *Environmental Humanities and Theologies* shows that some of these ways are problematic. It also provides alternative ways that value both materiality and spirituality.

Beginning with an environmentally friendly reading of the biblical story of creation, *Environmental Humanities and Theologies* goes on to discuss in succeeding chapters the environmental theology of wetlands, dragons and watery monsters (including crocodiles and alligators) in the Bible and literature. It then gives a critical reading of the environmental theology of the biblical book of *Psalms*. Theological concepts are found in the works of English writers of detective and devotional stories and novels, American nature writers and European Jewish writers (as succeeding chapters show). *Environmental Humanities and Theologies* concludes with an appreciation of Australian Aboriginal spirituality in the swamp serpent. It argues for the sacrality of marsh monsters and swamp serpents as figures of reverence and respect for living bio- and psycho-symbiotic livelihoods in bioregions of the living earth in the Symbiocene. This is the hoped-for age superseding the Anthropocene.

Environmental Humanities and Theologies is aimed at those who have little or no knowledge of how theology underlies much thinking and writing about 'the environment' and who are looking for ways of thinking about, being and living with the earth that respect and value both spirituality and materiality. It is a new text nurturing sacrality for the Symbiocene.

Rod Giblett is Honorary Associate Professor of Environmental Humanities in the School of Communication and Creative Arts, Deakin University, Australia. He is the author of many books in the environmental humanities, including *People and Places of Nature and Culture* (2011) and most recently, *Cities and Wetlands: The Return of the Repressed in Nature and Culture* (2016).

Routledge Environmental Humanities

Series editors: Iain McCalman and Libby Robin

The *Routledge Environmental Humanities* series is an original and inspiring venture recognising that today's world agricultural and water crises, ocean pollution and resource depletion, global warming from greenhouse gases, urban sprawl, overpopulation, food insecurity and environmental justice are all *crises of culture*.

The reality of understanding and finding adaptive solutions to our present and future environmental challenges has shifted the epicentre of environmental studies away from an exclusively scientific and technological framework to one that depends on the human-focused disciplines and ideas of the humanities and allied social sciences.

We thus welcome book proposals from all humanities and social sciences disciplines for an inclusive and interdisciplinary series. We favour manuscripts aimed at an international readership and written in a lively and accessible style. The readership comprises scholars and students from the humanities and social sciences and thoughtful readers concerned about the human dimensions of environmental change.

Environmental Humanities and Theologies

Ecoculture, Literature and the Bible

Rod Giblett

Routledge
Taylor & Francis Group
LONDON AND NEW YORK

earthscan
from Routledge

First published 2018 by Routledge

2 Park Square, Milton Park, Abingdon, Oxfordshire OX14 4RN
52 Vanderbilt Avenue, New York, NY 10017

Routledge is an imprint of the Taylor & Francis Group, an informa business

First issued in paperback 2020

British Library Cataloguing-in-Publication Data
A catalogue record for this book is available from the British Library

Library of Congress Cataloging-in-Publication Data
Names: Giblett, Rodney James, author.
Title: Environmental humanities and theologies : ecoculture, literature and the Bible / Rod Giblett.
Description: Abingdon, Oxon ; New York, NY : Routledge, 2018. | Series: Routledge environmental humanities | Includes bibliographical references and index.
Identifiers: LCCN 2017060130 (print) | LCCN 2018013642 (ebook) | ISBN 9781351124102 (eBook) | ISBN 9780815357643 (hardback)
Subjects: LCSH: Ecotheology. | Human ecology—Religious aspects—Christianity. | Nature in the Bible.
Classification: LCC BT695.5 (ebook) | LCC BT695.5 .G53 2018 (print) | DDC 261.8/8—dc23
LC record available at https://lccn.loc.gov/2017060130

ISBN: 978-0-8153-5764-3 (hbk)
ISBN: 978-0-367-49759-0 (pbk)

Typeset in Times New Roman
by Swales & Willis Ltd, Exeter, Devon, UK

Dedicated to Noel

Contents

Acknowledgments

The dedication of this book to Noel is both to Dr Noel Vose *in memoriam* and to my brother named after Dr Vose, both of whom have shaped my life in different ways at different times in relation to theology and spirituality. They bear no responsibility for what follows.

I am grateful to the Dalton McCaughey Theological Library in the University of Divinity, Melbourne for granting me access to some books by Habel and Westermann about *Genesis* cited in chapter 1.

For chapter 2, I am grateful to John Ryan for re-igniting my interest in Tolkien's and *Beowulf*'s take on wetlands. I first discussed *Beowulf* in *Postmodern Wetlands.* In an early draft of that book I also discussed the Dead Marshes in Tolkien's *The Lord of the Rings*, but this was later cut for reasons of length and lost. Placing the Dead Marshes in close context with *Beowulf* and my work on the World War I has enabled me to produce a richer reading of Tolkien's work. I am also grateful to John for passing on the reference to *The Silmarillion* and to two anonymous reviewers of *Green Letters* where this chapter was first published for their helpful comments, criticisms and suggestions that enabled me to broaden my range of references and sharpen the argument. I am grateful to Sandra Giblett for drawing my attention to Diana Gabaldon's *An Echo in the Bone* and her account of 'The Great Dismal' Swamp.

An earlier and shorter version of chapter 2 was published as: 'Theology of wetlands: Tolkien and *Beowulf* on marshes and their monsters', *Green Letters: Studies in Ecocriticism, 19* (2) 2015, pp. 132–143, © Association for the Study of Literature and the Environmental (ASLE UK-1), reprinted by permission of Taylor & Francis Ltd, www.tandfonline.com on behalf of Association for the Study of Literature and the Environmental (ASLE UK-1)

I am also grateful to three anonymous reviewers of the initial book proposal and a sample chapter for the Routledge 'Environmental Humanities' series for their encouraging and critical comments and helpful suggestions. They have certainly helped me to clarify the aims of the book, define its scope, improve its cohesion, strengthen the argument and include a discussion of some important references that I had overlooked, as I was a relative novice in ecotheology. Lurking quietly in the background during the process were Iain McCalman and Libby Robin, the editors for the series, who provided generous support at crucial moments, for which I am grateful. As ever, Sandra Giblett was there for me too.

Frontispiece:

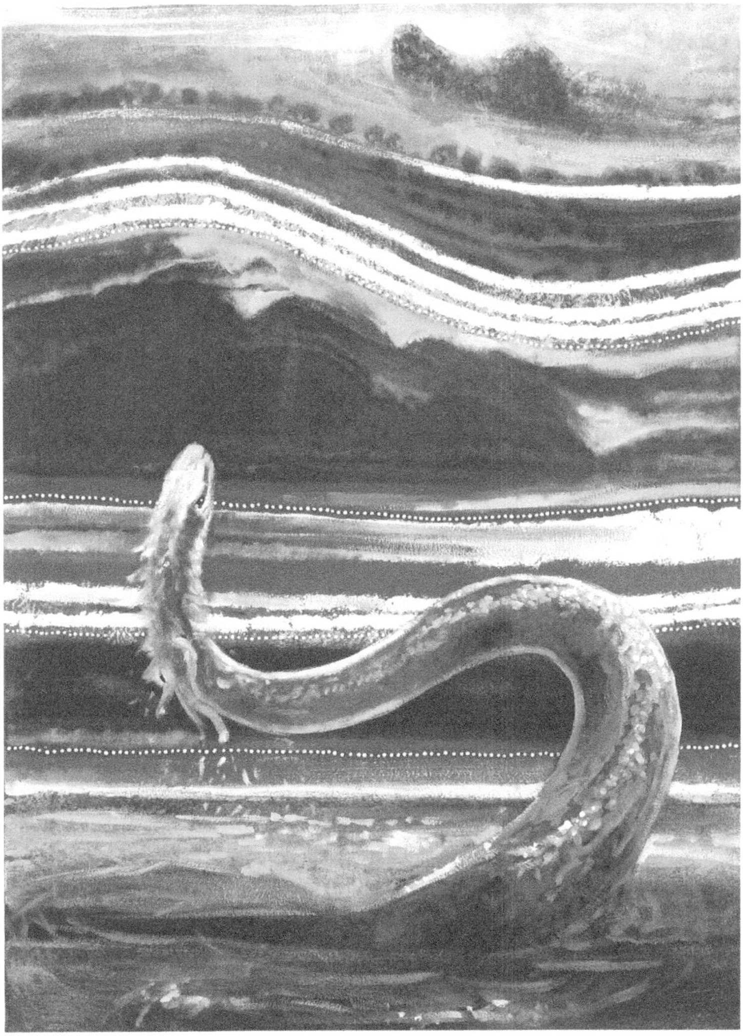

Shane Pickett/Meeyakba (1957–2009), *The Waagle and Yondock Story*, 2004, acrylic paint on canvas, 126 x 95 cm, City of Fremantle Art Collection, no. 951

Part I

Sacred earth and evil beings

1 In the beginning — was the wetland

In the beginning was the wetland. The earth and the water were without form and were chaotic, and darkness and light moved over the face and body of the earth and water. Earth and water were together one wet land. This was the first act of creation, the first coming into being, when the world was wetland, and the wetland was womb from which all later life sprung. Including human life. The wetland is not only the womb, but the womb is also a wet land, a slimy swamp of embryonic life. The wetland is not only the womb out of which all life came, but also the tomb into which all life dies and from which new life is (re)born. This was the world before the Fall, when the world was good and without evil, before the swamp and the marsh became places of darkness, disease and death, home alone to grotesque monsters lurking in the uncanny depths of their murky waters evoking horror and fascination in any who should have been unfortunate enough to stumble upon them.

Genesis

Is this rewriting of *Genesis* 1: 1–2 heresy or good theology? Or is it a bit of both? It is not a translation, nor a paraphrase of existing English translation, nor does it have a biblical basis or a traditional theological warrant. It is a rewriting of *Genesis* 1: 1–2 from a conservationist and environmentalist point of view. This is its only warrant (for an earlier discussion along these lines see Giblett, 1996, pp. 142–143). *Genesis* 1: 2 is translated into English in the official versions of the Bible as 'the earth was without form, and void; and darkness was upon the face of the deep: and the Spirit of God moved upon the face of the waters', or as 'the earth was formless and empty, darkness was over the surface of the deep, and the Spirit of God was hovering over the waters'.

For Norman Habel in his ecological reading of the biblical creation story:

> The image projected in *Genesis* 1: 2 is one of total darkness. It is a picture of a mysterious primordial realm where nothing is visible. There is no negative verdict on the darkness, no designation of the darkness as evil, sinister, or forbidding [. . .] The second component of the primordial state is 'the

> waters' [. . .] No negative verdict is pronounced on these waters – there is no indication that they are threatening powers or turbulent forces of chaos. They are part of the benign dormant primordial order.
>
> (Habel, 2000, pp. 36–37)

On Habel's reading of *Genesis* 1: 2, wetlands and other waters are not places of darkness in a pejorative sense of being evil, sinister, or forbidding, nor are their waters threatening powers or turbulent forces of chaos. They are part of the benign dormant primordial order, but waters (and wetlands by implication) are not always and only benign. They are part of the dormant primordial slime. Waters can be threatening powers and turbulent forces of chaos. Westermann (1984, p. 105) comments that *tehom* 'in the Old Testament has no other meaning than the deep or the waters of the deep. In can mean that which blesses or destroys just as can water'. Water is creative and destructive, life-giving and death-dealing.

Habel is indebted to Westermann when he embraces the first sentence in supporting his reading of *tehom*, but he does not accept the second one. Habel's ecological reading of the biblical creation story projects 'no negative verdict' on darkness and waters, but this is not wise ecologically and culturally. His reading also demonstrates the difficulty of ascribing no values to darkness and waters, and to being neutral about both, for which *Genesis* 1: 2 seems to be striving, or to at least acknowledging that waters and darkness have a negative (but not evil) aspect. Westermann (1984, p. 104) also comments that *chashak*, translated as darkness, is 'the opposite of creation. Darkness is not to be understood as a phenomenon of nature but rather as something sinister', as threatening (and Habel seems to be alluding to, and repudiating, this interpretation) and as 'the darkness of chaos'. Or, the chaos of darkness. On this reading, the darkness is certainly not evil, as evil has not been introduced into the creation and so the creation is good. On this reading also, the waters are not chaotic, but darkness is. Darkness is chaotic. Creation comes out of chaotic darkness, and not out of chaotic waters. When God divides land from water, He is not creating order out of chaos, but making dry land out of wet land, creating an amenable place for patriarchal pastoralism and agriculture to take place. He is also making a place suitable for cities, usually made of solid materials, to be located.

A decade later Habel (2011, p. 30) is more willing to embrace a reading of *Genesis* 1: 2 even more in line with my rewriting of it in wetland-friendly terms. The breath of God hovering over the primordial waters 'suggests an image of parental nurture rather than primal disturbance'. Habel is also willing to locate himself 'within the primordial habitat of *Erets* [earth] and identify with her as a character in the primal scene'. More precisely, she is obviously gendered as a feminine character and the mother is the usual provider of parental nurture, certainly exclusively with breast milk. Habel (2011, p. 30) goes on to comment that 'the imagery of this habitat or scene suggests an embryonic figure without the form and fertility associated with the land mass called *Erets.* The scenario suggests a primal womb embracing the unborn *Erets.*' Yet the landmass *Erets* is not, and can never be, an exclusively dry land mass. It is not fertile per se.

When *Erets* assumes form and fertility, it is a wet land mass as well as a dry land one. The *Erets* is a land and water mass.

In a later section headed 'The Birth Metaphor' Habel comments further that:

> the implication of this reading that a womb/birth metaphor lies behind the imagery for the setting and appearance of *Erets* on day three may seem surprising, given the tendency of many interpreters to view *tehom* and the waters as evidence of primal chaos. That *Erets* is viewed as a mother in some biblical passages is well-known (*Psalm* 139: 13–15).
>
> (Habel, 2001, p. 31)

Yet *Erets* appeared on day one when God created the earth, and not on day three when God separated the land and waters. In the light of Habel's earlier readings of *tehom* and the waters *not* as primal chaos, the implication of his reading is not that surprising.

What Habel does not address in his reading of *Genesis* 1 and in his repudiation of primal chaos and the old *Chaoskampf* reading of the biblical account of creation is that this older reading of the first book in the Bible is important culturally, especially for its associations with monsters and dragons. Habel, in particular, repudiates the connection of '*tehom* (deep)' with '*Tiamat*, the chaos waters deity found in [. . .] Babylonian [. . .] myth' (Habel, 2011, p. 29). More precisely, *Tiamat* in Babylonian myth is for Westermann (1984, pp. 29 and 31) 'the primeval monster of chaos' and 'chaos' is, or is interchangeable with, 'a dragon'. John Milton in the seventeenth century did not repudiate the connection and was not successful in keeping the primeval monster of mythological Babylonian chaos at bay (as we will see shortly).

Irrespective of the biblical warrant for doing so, or for reading *Genesis* 1: 2 in terms of primal chaos, the association between the monstrous, the serpent, primal chaos, especially construed as their swampy habitat in hell, is a persistent trope in western European thinking and writing about creation, water and wetlands. When it comes to creation, water and wetlands, theological thinking and writing not only gets mixed up with Babylonian mythology, but also with Greek philosophy, in a syncretistic mélange (and John Milton's *Paradise Lost* is also a case in point as we will see shortly after a brief excursion into a discussion of the four elements to provide the background).

Four elements

In European philosophy going back at least to Aristotle there are four elements: earth, air, fire and water.[1] The four elements of earth, air, fire and water are combinations of the four primary qualities of hot, cold, wet and dry. According to this philosophy, water is the combination of the qualities of cold and wet; fire is the combination of the qualities of hot and dry; earth is the combination of the qualities of cold and dry; and air is the combination of the qualities of hot and wet, as shown in Figure 1.1.

	HOT	COLD
DRY	FIRE	EARTH
WET	AIR	WATER

Figure 1.1 The Four Elements

Rather than being primary matter themselves, the first substances, the elements are products of primary qualities. Qualities on this view precede quantity; process precedes products.

Of course, in an age of global warming water can not only be wet and cold, but also wet and hot (and increasingly hot in the age of rising sea temperatures creating more frequent and intense hurricanes). Earth can not only be dry and cold, but also dry and hot (and increasingly hot in the age of more frequent and intense bush and forest fires). Air can not only be wet and hot (and increasingly hot with rising air temperatures and each year hotter than the last), but also dry and cold. But these are aberrant combinations according to the philosophy of the four elements. They are ruled out of the order of things based on a hard and fast distinction between the qualities and between the elements based on their proper combination of the primary qualities. Wet and hot air (and increasingly so), such as in the humid tropics and sub-tropics, and increasingly so in 'Mediterranean' and temperate climates, is generally regarded as uncomfortable. Cold and wet earth, such as in temperate wetlands, or hot and wet earth, such as in tropical and sub-tropical wetlands, or hot and dry earth everywhere else (and increasingly so) have been regarded as unacceptable.

Watery earth that mixes the two elements of earth and water has been regarded as aberrant, as dirt, as matter out of place, to use Mary Douglas's terms (Douglas, 1966, p. 35). Wetlands have been considered as elemental aberrations, though these are creative and productive mixing of the elements in nature's womb of mud (earth and water). Increasingly, elemental aberrations in the age of global warming are the new norm, though these are destructive mixings of the elements in culture's tomb of crud (including mud). Increasingly in the age of global warming/climate change there is more and more dirt, more and more matter out of place, requiring bigger and bigger clean-up operations, costing more and more money. Thinking with the qualities and the elements helps to identify and sort through these different sorts of mixtures and aberrations with their intertwining of nature and culture, and to living with the earth in a better, more symbiotic way by, for a start, drastically reducing greenhouse gas emissions (dirty air).

Nature's womb

John Milton's *Paradise Lost* will suffice for the time being in this chapter as one case in point, as one proof text, of aberrant wetlands under the philosophy of the four elements. Reading Milton's biblical epic poem ecocritically also provides an opportunity to rewrite ecocreatively some sections of *Paradise Lost* to showcase wetlands as a fertile mix of qualities and elements. Milton followed Aristotle in suit by only admitting the possibility of the mixture of four elementary qualities into what he called 'the cumbrous elements, earth, flood, air, fire' (III, 715):

> [. . .] ye elements, the eldest birth
> Of Nature's womb, that in quaternion run
> Perpetual circle, multiform, and mix
> And nourish all things, let your ceaseless change
> Vary to our great Maker still new praise.
>
> (V, 180–184)

The mixing of qualities in 'Nature's womb' to procreate the elements does indeed nourish all things as Milton indicates, but not only the four mixtures of qualities into the four elements that Aristotle and Milton permit. Indeed, the two aberrant mixtures of wet and dry, hot and cold produce the temperate and tropical wetland which are far more productive than both dry land and cold water. It is the wet land and warm water which give birth to life and which are the womb of the world.

Rather than the womb of the world, wetlands for Milton are the tomb in the underworld. In *Paradise Lost* Milton saw the Styx, one of the rivers of the classical underworld and for Dante a slimy swamp in the fifth circle of the Inferno (as we will see in a later chapter), as 'the burning lake', as 'abhorrèd' and as 'the flood of deadly hate' (II, 576, 577). In the same book of *Paradise Lost* he elaborates later in more detail on:

> [. . .] lakes, fens, bogs, dens and shades of death,
> A universe of death, which God by curse
> Created evil, for evil only good,
> Where all life dies, death lives, and Nature breeds,
> Perverse, all monstrous, all prodigious things,
> Abominable, inutterable, and worse
> Than fables yet have feigned, or fear conceived,
> Gorgons and Hydras, and Chimeras dire.
>
> (II, 621–628)

This view of hell as swamp and swamp as hell persists through the nineteenth century and well into the twentieth century, such as in J. R. R. Tolkien's Dead Marshes in *The Lord of the Rings* (as we will see in the following chapter; see also Giblett, 1996, p. 140).

Milton's pejorative theology of wetlands needs rewriting in environmentally friendly terms as places of both life and death:

> Lakes, fens, bogs, dens and spirits of life,
> A universe of life, which God by love

Created good, only for good, not evil
Where all life dies, new life lives, and Nature breeds,
Promiscuous, all maternal, all prodigious things,
Fascinating, horrifying, and better
Than fables yet have feigned, or fear conceived,
Marsh monsters, swamp serpents and dragons fiery.

These monstrous beings embody the mixing of the four elements of earth, air, fire and water (as we will see in later chapters, including in relation to Satan in *Paradise Lost*).

Later in the same book of *Paradise Lost* Milton describes in more detail the watery world and womb of the first day of creation that morphs into Hell:

[. . .] this wild abyss,
The womb of Nature and perhaps her grave,
Of neither sea, nor shore, nor air, nor fire,
But all these in their pregnant causes mixed
Confus'dly, and which thus must ever fight,
Unless th' Almighty Maker them ordain
His dark materials to create more worlds,
Into this wild abyss the wary Fiend
Stood on the brink of Hell.

(II, 910–918)

Milton's Satan, referred to here as 'the wary Fiend', is a swamp serpent and marsh monster, or more precisely a miasmatic marsh monster, an elemental aberration who mixes earth, air, fire and water (as we will see in chapter 3).

Milton's philosophical theology of wetlands again needs rewriting in environmentally friendly terms as fertile, liminal places mixing creatively the four elements of earth, air, fire and water:

This wild swamp,
The womb of Nature and perchance her tomb,
Of foursome sea, and shore, and air, and fire,
With all these in their pregnant elements mixed
Creatively, and which thus must ever in concord,
That the All-making Mater them ordain
Her dank materials to create more worlds,
Into this wild swamp the wary friend
Stood on the brink of Heaven.

And not 'on the brink of Hell', for this is God's first and best work in the beginning on the first day of creation. Without it no other day could ever begin, could ever dawn. Rather than being hell, the wetland is the closest thing to heaven because it was created next, right after heaven. When one goes to the wetland one goes back to the first day of creation when it springs fresh and new from the hand and word of God, when the breath of God broods over the earth as wetland, like a hovering marsh harrier not like a malevolent ghost.

God, for Milton, brings order out of the wetland chaos, out of 'the vast immeasurable abyss/Outrageous as a sea, dark, wasteful, wild' (VII, 211,2):

> [. . .] on the wat'ry calm
> His brooding wings the Spirit of God outspread,
> And vital virtue infused, and vital warmth
> Throughout the fluid mass, but downward purged
> The black tartareous cold infernal dregs
> Adverse to life [. . .]
>
> (VII, 234–239)

Hell as swamp was not just any part of hell for Milton, but the lowest part of the classical Greek underworld called Tartarus. Milton regards water as lifeless until the spirit of God breathes life into it. Milton's theology entails sublimation of the materiality and immanent spirituality of earth and water into the transcendental spirituality of God.

Milton's theology has little or no biblical basis or warrant in this regard, but comes from Western philosophical idealism that separates spirit and matter. It also valorizes spirit as life-giving force over matter as lifeless substance until spirit infuses it with life (as Milton says). This kind of syncretism also characterizes ecotheology, such as the distinction between spirit and matter not found in the Bible. Even though breath and spirit are the same word in Hebrew (and Greek), Theodore Hiebert (2008, p. 18) argues that 'it is improbable that the word "spirit" is ever a viable interpretation for the biblical Hebrew רוּחַ'. He translates רוּחַ as breath, air, the atmosphere, God's breath and 'the first sacred thing' — and not as spirit. He adds that to interpret the biblical Hebrew רוּחַ as spirit 'can only seriously distort the biblical perspective' as the Hebrew Bible does not separate out spirituality and materiality, but sees them as interconnected. Hiebert argues that the 'post-biblical theological dualism' of matter and spirit comes from Western philosophical idealism originating with Plato. In addition, the concept of matter itself comes from Western philosophy more generally, and matter as made up of indivisible particles comes from Western philosophical atomism originating with Democritus (Lloyd and Sivin, 2002, p. 150).

Milton's derogatory theology of creation again needs rewriting in sacral wetland-friendly terms:

> On the wat'ry calm
> Her brooding wings the breath of wind outspread,
> And vital virtue suffused, and vital warmth
> Throughout the fluid mess, so outward splurged
> Her black, warm, slimy dank materials
> Conduce for life.

A wetland-friendly philosophy and theology of creation entails desublimation of solid matter (earth) into the immanent spirituality (breath, air; warm, fiery) of liquid materiality (water).

Kelly Barnhill's best-selling *The Girl who Drank the Moon* provides such a wetland-friendly philosophy and theology of creation complete with a monster:

> In the beginning, there was only Bog, and Bog, and Bog. There were no people. There were no fish. There were no birds or beasts or mountains or forest or sky.
>
> The Bog was everything, and everything was the Bog.
>
> The muck of the Bog ran from one edge of reality to the other. It curved and warbled through time. There were no words; there was no learning; there was no music or poetry or thought. There were just the sigh of the Bog and the quake of the Bog and the endless rustle of the reeds.
>
> But the Bog was lonely. It wanted eyes with which to see the world. It wanted a strong back with which to carry itself from place to place. It wanted legs to walk and hands to touch and a mouth that could sing.
>
> And so the Bog created a Body: a great Beast that walked out of the Bog on its own strong, boggy legs. The Beast was the Bog, and the Bog was the Beast. The Beast loved the Bog and the Bog loved the Beast, just as a person loves the image of himself in a quiet pond of water, and looks upon it with tenderness. The Beast's chest was full of warm and life-giving compassion. He felt the shine of love radiating outward. And the Beast wanted words to explain how he felt.
>
> And so there were words.
>
> And the Beast wanted those fit together just so, to explain his meaning. He opened his mouth and a poem came out.
>
> 'Round and yellow, yellow and round,' the Beast said, and the sun was born, hanging just overhead.
>
> 'Blue and white and black and gray and a burst of color at dawn,' the Beast said. And the sky was born.
>
> 'The creak of wood and the softness of moss and the rustle and whisper of green and green and green,' the Beast sang. And there were forests.
>
> Everything you see, everything you know, was called into being by the Bog. The Bog loves us and we love it.
>
> (Barnhill, 2017, pp. 83–84)

The Bog, like all wetlands, is a liminal zone between earth and water, between solid and liquid. This liminal zone is the most fertile zone on the planet, where new life begins again and again. This fertile, liminal zone is also primeval slime. Like the wetland, slime is between the solid and the liquid. The four elements of earth, air, fire and water are thus related to four primary categories of substance: earth is solid; fire is hot (duh); air is gaseous; and water is liquid. Frozen or solid water is of little use except for walking on, or sledding on, or hunting on, or keeping drinks and other things cool. Slime is between the solid and liquid. Slime is horrific for Jean-Paul Sartre, or abject for Julia Kristeva (see Giblett, 1996, especially chapter 2). The sublime chemically is the transformation of a solid into a gas; the sublime is between the solid and the gaseous, or earth and air. Culturally and psychologically the sublime and sublimation are the displacement of the lower into the upper, of the grotesque lower bodily, monstrous marshy and swampy strata into the sublime upper mental and mountainous heights of abstraction and theory.

Water

Water has been associated for a long time with monsters such as crocodiles or sharks or imaginary creatures of bogs and marshes (as we will see in later chapters). Marsh monsters, swamp serpents and fiery dragons have inhabited the human imagination and water bodies or desert wastes for a long time. These creatures, rather than beings in themselves, are emanations and expressions of the monstrous mixing of the elements and their qualities. Water can be monstrous, as the current or recent flooding somewhere in the world shows. Evoking the monstrous qualities and properties of water helps humans to learn to respect water, if not to sometimes fear it. Water is not only in swamps, but can swamp. Swamp is not just noun, but also verb. Water knows no bounds, except for those created for it by land and by humans. Water drowns fire, can drown earth and can drown humans. In drowning, water overcomes the boundary between the inside and outside of the body.

The monstrous qualities and properties of water remind humans of the water without and the water within us. Water for the pre-Socratic philosopher Thales of Miletus is 'the source of all things' (cited by Giblett, 1996, p. 144). Water is the primary element. In biblical terms (as we have seen), in the beginning was water, then there was wet land. Water for psychoanalytic philosopher Gaston Bachelard (1983, p. 6) is the foremost component. Of what, one may ask? Of everything. Or of most living beings. Water is the foremost component of the planet (about 71% of the Earth's surface is water-covered). Water is also the foremost component of the human body with about the same percentage. Our bodies are predominantly water. Whether we modern urban humans are predominantly water is another question as we tend to life in our minds and be separated from our bodies. Our waters are primarily in the depths of our bodies, whereas the earth's waters are primarily on its surface, but also in its depths in aquifers, or underground rivers and streams.

The monstrous qualities and properties of water remind humans of the water in which we were conceived. Human bodies are conceived in water and grow in the waters of the womb. The monstrous qualities of water remind us of the first maternal liquid which we drank. Water is maternal and water is primal milk. For Bachelard (1983, pp. 117–118; see also Giblett, 1996, pp. 175–176) 'all water is a kind of milk'. Conversely, mother's milk (whether it comes from a human or other animal) is primal water. Milk is typically white, but water is white, black and coloured. White or clear water is life-giving as drinking water, whereas black water is death-dealing, but it is also life-giving as a womb for the creation of new life. Water is not neutral; the colour of water matters; the colour of water is cross-culturally coded between Australian Aboriginal people and their settler counterparts (see Giblett, 2007; 2009, chapter 9). The philosophy of water is cross-culturally coded too. Water for traditional Chinese Taoists exemplifies the soft and yielding power of the Tao, or the Way (see Giblett, 2013, chapter 17).

Sacred earth and evil beings

The association between the monstrous, the dragon and primal chaos, especially construed as their habitat, has a biblical warrant in rereading *Genesis* 1: 2 without earlier Babylonian mythological and later Aristotelian philosophical accretions and associations. Westermann (1984, p. 76) translates *tohu wabohu* in *Genesis* 1: 2 as 'desert waste', usually translated as 'without form, and void', or as 'formless and empty'. Westermann (1984, p. 103) comments later that *tohu wabohu* should be 'understood as something gruesome'. Westermann's translation and commentary is highly pertinent culturally to the Old English poem *Beowulf* dating from the eighth century of the Common Era, its depiction of the dragon and the description of the surrounds of his home. The dragon's mound is described in Seamus Heaney's translation as surrounded by 'that desert waste' (*Beowulf*, 1999, p. 73). The dragon in biblical terms is a creature of the desert waste of 'primeval chaos', or 'the formless and empty', created before the animals. He is a survivor of, or throwback to, the first day of creation just after God created the heavens and the earth, but before He divided the heavens, waters and earth from each other, and light from darkness. The dragon is a creature of primeval chaos; in a sense, he *is* primeval chaos, or certainly a manifestation of it. He is also gruesome, as described in J. R. R. Tolkien's translation of *Beowulf* (Tolkien, 2014, p. 95).

The dragon is created before the Fall whereas other monsters, such as Grendel and his mother in *Beowulf*, are created after the Fall. Chapter two considers these two monsters within the broader context of the theology of wetlands and their monsters, including in Tolkien's work. From the Old English *Beowulf* to Tolkien's *The Lord of the Rings* and his critical and scholarly work on Old and Middle English literature, wetlands and marsh monsters get bad press. *Beowulf* interpolates several verses into the biblical book of *Genesis* that explain the propagation of monsters and justify regarding wetlands as evil, pagan places. Tolkien developed this view of marshes and marsh monsters with his experience of the horrific wet wasteland of the Western Front in World War I and applied it to the 'Dead Marshes'. He is what I call 'placist' as he ascribes moral qualities to a place undeserving of such opprobrium and gives it derogatory qualities in the lower psychopathological register. This chapter traces critically the representation of wetlands and marsh monsters in both texts and argues for a more environmentally friendly view of wetlands as places of life and death combined and embodied in the figure of the monster.

From the Old English *Beowulf* through the legend of St George and the dragon and Malory's *Morte d'Arthur* to Tolkien's *The Hobbit*, dragons also get bad press. A monstrous, draconian serpent also makes an appearance as Satan in Milton's *Paradise Lost* and in fairy tales. Chapter three traces critically the representation of dragons in all these texts. It begins by taking up *Beowulf* again in which the dragon is an earth dragon, as is Smaug in *The Hobbit*, whereas the dragon in the legend of St George and in *Paradise Lost* is a swamp dragon, as is Satan in Milton's *Paradise Lost* too. This chapter argues for a more environmentally friendly view of swamp dragons as manifestations of the fertility of the earth to live with and use sacrally and symbiotically, rather than to kill. Earth dragons wreaking fiery vengeance on humans for stealing from the earth can also be read as a climate change allegory. More intense bush and forest fires in the age of global warming are the earth's vengeance for greedy mining and carboniferous capitalism (see Giblett, 2018b).

Another watery monster makes an appearance in the Bible in the form of Leviathan, an ambiguous creature who may be a whale or a crocodile. Crocodiles also make regular appearances in northern Australia and stories about them are invariably couched in biblical terms as monsters of the ambiguous zone between land and water. Chapter four traces critically the representation of them in the Bible and literature and argues for a more environmentally friendly view of them as a companion species on the home planet earth. Rather than humans invading and destroying crocodile habitat and demonizing them, humans could learn to co-exist and co-inhabit the earth, but not in the same places. Rather than regarding marsh monsters and swamp serpents with horror and demonizing them as evil beings, they should be respected and revered as fellow creatures of the sacred earth.

Evil beings and the sacred earth appear in the biblical book of *Psalms*, probably the most lyrical book of the Bible. C. S. Lewis, the twentieth-century English literary scholar, children's and sci-fi author and popular Christian writer, at least thought that it contains some of the best lyrical poetry in the world, including about the earth and its benign and beneficent places and features. It paints a bucolic picture of pastoral parts of the earth and some of its creatures frolicking playfully. It also contains some problematic environmental theology in which the earth is objectified as an object of God's ownership.

In the early 2000s the Earth Bible Team (made up of believers writing for believers and led by Norman Habel) attempted to retrieve an environmentally friendly Bible in terms of an inter-subjective relationship between God, humans and earth. Retrieval is coincidentally one of Habel's words used later to describe the Earth Bible Team project (Habel, 2008, p. 5). Retrieve is also one of the rival Earth Project's term to describe the response of Christian ecotheology to 'ecological destruction and environmental injustices' (Conradie *et al*, 2014, p. 1). Retrieve and retrieval imply that 'the ecological wisdom embedded in the Christian tradition' (as Conradie and his co-authors and contributors put it) has been lost or buried and is just there waiting to be found, rescued from oblivion and allowed to respond to 'ecological destruction and environmental injustices', though to what end or purpose this response is made is not specified. Presumably this response is designed to halt ecological destruction and further environmental justice, invariably tied up with social justice.

The 'Christian tradition' implies that ecological wisdom is not just to be retrieved from the Bible. Ecotheology does so from the Church fathers and later theologians, such as Thomas Aquinas and Martin Luther, but not from nature and other writers writing in the Christian tradition, such as John Muir, whose writings are often overlooked by theologians. Retrieve and retrieval also imply that the projects of ecotheology tend to be highly selective and overlook or ignore and not critique the ecological folly in the Bible, such as *Leviticus* 11: 41 that 'every swarming thing [. . .] is an abomination' (discussed in greater detail in later chapters). It is also tends to overlook or ignore and not critique the ecological folly in the Christian tradition (including from *Beowulf* through *Pilgrim's Progress* and John Muir's nature writings to *The Lord of the Rings*). Both ecological wisdom and ecological folly can be found in Muir's work. Appreciating the former and critiquing the latter are crucial for environmental theology in which both need to be acknowledged and identified. Not to do so would be to place ecological wisdom in danger from ecological folly and whitewash the latter with the former.

The projects of retrieval by ecotheology are also problematic as they uncritically reproduce concepts from Western philosophy, such as inter-subjectivity, rather than from Christian theology, in their readings of the Bible, including *Genesis* and the *Psalms*. In this regard, they are following in the tried and trusted syncretistic footsteps of Milton's *Paradise Lost* and his revisionist rereading of *Genesis*. Chapter five discusses and critiques the environmental theology of the *Psalms* and argues for a more environmentally friendly theology of an inter-corporeal relationship between the sacred body of the earth and the sacred human body, nurturing mutuality and symbiosis with the earth and its creatures, especially with water-beings who generally get short shrift in the *Psalms* (as chapter five shows) and the Old Testament in general (as chapters one to four show).

Theologies of times and places

Theological concepts and tropes, some drawn from the *Psalms* (such as the valley of the shadow of death), permeate Western culture, including its literature, again often overlooked by theologians. John Bunyan's *Pilgrim's Progress*, first published in the seventeenth century, is the second-most published book in English after the Bible, but it contains some environmentally unfriendly theology, especially when it comes to dragons, the slough of despond and the valley of the shadow of death. All three are theologized and moralized as evil. The slough of despond has no biblical warrant, whereas dragons and the valley of the shadow of death do (including in the *Psalms* as indicated previously and as discussed in chapter five).

The association between sloughs and despondency has its roots in the theory of the elements and the humours and its flowers in Robert Burton's *Anatomy of Melancholy* and Bunyan's *Pilgrim's Progress.* Later uses of these tropes, such as in C. S. Lewis' *The Pilgrim's Regress*, follow the Psalmists' lead in theologizing dragons as evil beings and their places of abode as evil. Nineteenth-century fiction, especially detective fiction, draws on the slough of despond and the valley of the shadow of death as tropes for evil places. These writers, like Tolkien, are 'placist' in that they ascribe characteristics to a (human-made) place in the city, such as the slums, which were previously ascribed to a place not made with human hands, or made with ancient human hands (jungle, abyss, nether-land, swamps, etc.), such as darkness, decomposition and disease. Chapter six discusses and critiques the environmental theology of the elements, the humours, the slough of despond, dragons and the valley of the shadow of death. It argues for a more environmentally friendly theology of wetlands, monsters, death and wastelands as places and beings of both life and death.

Theological concepts and tropes also permeate the diasporas of Western culture in the colonies of the European colonizers. Some settlers of the United States saw their initial trans-Atlantic and later trans-continental migration in biblical terms, as a mission of God's chosen people crossing the frontier into the wilderness, finding the promised land beyond and fulfilling their 'manifest destiny' to institute the kingdom of God on earth. Many biblical concepts and terms were used in early settler discussions of the American landscape and in the development of national parks. John Muir was the founding President of the Sierra Club. He followed in the footsteps of the prophetic founder of the American Conservation Movement, Henry David Thoreau, who wrote the old testament of conservation in *Walden* and his other writings. He is also arguably the 'Patron Saint of Swamps'.

As 'John the Baptist' crying in the wilderness, Muir prepared the way for the Jesus of Aldo Leopold and his saving gospel of a land ethic and conservation aesthetic in the new testament of *A Sand County Almanac.*[2] Muir expressed his nature aesthetic of mountains in biblical and Neolithic terms, and of swamps in 'pagan' or Palaeolithic terms. Chapter seven discusses and critiques Muir's environmental theology and the gender politics of his contradictory ideas of nature divided between a stern and sublime Father God in the rocky mountains and a soft and smiling Mother Goddess in the slimy swamps. Muir's problematic theology brings into stark relief the contrast and contest between the two paradigms (ways of thinking, being, living and making meanings) of patriarchy and matriarchy, or more precisely, filiarchy and matrifocality, or simply mastery and mutuality.[3]

Destruction and creation are natural processes. Muir saw both occurring in swamps and mountains. They are also cultural and theological processes, including the creation and destruction of the cultural environment of cities, often couched in biblical terms of the destruction of Sodom and Gomorrah. Chapter eight considers the biblical story in *Genesis* 19 of Lot's wife who was turned into a pillar of salt for looking back at God's fiery destruction of the sinful cities of Sodom and Gomorrah after God had warned her not to do so or she would be punished – as she indeed was. Being turned into a pillar of salt is a theological trope for those who are fixated on the past. The fiery destruction of the two sinful cities is also an apocalyptic trope for the end of the world, such as in writings about the firebombing and destruction of cities, and the end of the world, in the Second World War (as this chapter goes on to discuss).

The story of Lot's wife also resonates with the experience of exile and wandering of such writers as Walter Benjamin and Stefan Zweig, both of whom had a Jewish background. Both were Jews by birth, though not by faith, and both were regarded as Jews by the Nazis. Benjamin's writings are famously saturated or soaked with theology like used blotting paper, whereas Zweig's seem thoroughly secularized and bereft of theology like new blotting paper. Both writers were victims of Nazi Germany; both expressed their sense of exile in (different) ways that involved them looking back, but without becoming fixated on the past (and so heeding God's warning to Lot's wife), and, in Benjamin's case, invoking the Angel of History who looks back at destruction with her back turned toward the future.

By contrast with Benjamin and Zweig, Henry David Thoreau calls for, lives and exemplifies a non-exilic experience of being at home in one's body, time and place. Chapter eight goes on to discuss Thoreau's life and work in these terms. Following in Benjamin's footsteps, the body, time and place are presided over by the Angels of History, Geography and Corporeality. They acknowledge and respect processes of creation and destruction in an environmentally friendly theology of what Thoreau as the prophet of the old testament of conservation and as the 'Patron Saint of Swamps' called the 'quaking zone' of the body in its nether regions and of the earth in its grotesque lower stratum of bogs and marshes in an inter-corporeal, multi-sensory relationship between the human body and the body of the earth.

A non-exilic experience of being at home in one's body, time and place is arguably a characteristic of indigenous cultures. The relationship between biblical environmental theology and indigenous cultures around the world has been fraught to say the least, no less so than in Australia where Aboriginal people have struggled to maintain their lands, native title, claims, cultures and beliefs in the face of colonizing dispossession and genocide. One figure where this relationship has played out is around the Rainbow Serpent or Rainbow Spirit. Chapter nine discusses the Rainbow Serpent in anthropology and the Rainbow Spirit in Aboriginal theology. This chapter (and the book) concludes by proposing the earthly sacrality of marsh monsters and swamp serpents as figures for a belief in, and as a way of living, bio- and psycho-symbiotic livelihoods in bioregions of the living earth in the symbiocene, the hoped-for age and paradigm superseding (in the sense of embracing and following) the Anthropocene.[4] I define, discuss and differentiate the symbiocene and the Anthropocene in greater detail in the final chapter.

Introducing *Environmental Humanities and Theologies*

Environmental Humanities and Theologies combines an environmental reading of the Bible and Christian-infused English literature (with a distinct emphasis on the monstrous and watery) with both a discussion of theologically-inflected American nature writers (in Muir and Thoreau) and European Jewish writers (in Benjamin and Zweig), and an appreciation for Australian Aboriginal spirituality in the swamp serpent (as portrayed in Shane Pickett's painting reproduced in the frontispiece and discussed in the final chapter). Like the Bible, it is a vital or eclectic mix of texts and has plenty of tangential aspects. But unlike the Bible, these aspects are part of a developing argument and a logical progression (one wonders if the Bible would be published if it were submitted for publication today), though of course the Bible has a chronological progression from the beginning of the world to the end of the world (which this book mimics to some extent with both books ending in hope, the Bible in hope for the return of Jesus the Messiah, this book in hope for messianic materiality and spirituality).

Like the Bible, *Environmental Humanities and Theologies* begins with creation in the book of *Genesis* before the fall from the Garden of Eden. It then: dwells on dragons and other monsters created before and after the Fall in *Genesis* and other books, such as *Beowulf*; passes through the book of *Psalms* and its resonances in John Bunyan's *Pilgrim's Progress* and other stories; explores Bunyan's departure from the *Psalms* in the slough of despond into which other writers have followed him; goes on to discuss critically John 'the Baptist' Muir's ecology and theology; addresses apocalypse, destruction of cities, the end of the world and the experience of exile; appreciates Benjamin's and Thoreau's theologies of being at home in one's time, place and body; invokes the attendant Angels of History, Geography and Corporeality; and concludes with and in hope in the past for the present and future with Aboriginal spirituality.

The first five chapters are gathered into the first part of the book, 'Sacred Earth and Evil Beings'. This part could be considered as a new old testament as it begins at the beginning with creation in the biblical book of *Genesis*, references the Old Testament of the Bible and ends with an ecocritical reading of the *Psalms*. The final four chapters constitute the second part of the book, 'Theologies of Times and Places'. This part could be considered as a new new testament as, like the Bible, it builds on, and refers to, the first part of *Environmental Humanities and Theologies* and to the Old Testament of the Bible, such as the story of Lot's wife. dragons also keep on cropping up and raising their ugly (or otherwise) heads in the second part in Bunyan's *Pilgrim's Progress* and C. S. Lewis's *The Pilgrim's Regress*. Similarly, the biblical Valley of the Shadow of Death in *Psalm* 23 reappears in the second part of the book in the progress of Bunyan's pilgrim from the city of destruction to the celestial city and in Fergus Hume's detective story, *Mystery of a Hansom Cab*, involving the hero descending into the dark underworld of the slums in the modern city. Also, as with the New Testament, new characters and concerns are presented in the second part with the introduction of John 'the Baptist' Muir.

Coming from within ecocultural studies and the transdisciplinary environmental humanities (crossing between history, philosophy, anthropology and literary studies, and engaging with environmental concerns and issues across these disciplines), as well as affirming psychoanalytic, postmodern and political (socialist and feminist) ecologies,[5] *Environmental Humanities and Theologies* respectfully engages in dialogue with theologians and creatively critiques theology. It is aimed in part at theologians unfamiliar with the use of theological concepts and tropes in literature (including nature writing). The author follows in the footsteps of Benjamin, Lewis and Tolkien who were primarily literary scholars and writers (like me) who wrote about biblical texts and theological themes and/or used theological concepts (as I am now doing in the present book). Lewis and Tolkien were not theologians (like me), but were Christians (unlike me), whereas Benjamin was neither (like me).

Environmental Humanities and Theologies is written by an ex-Christian who was born in Borneo of Australian Christian missionaries and who spent most of his childhood living in a Bible training institute where his father worked as the Registrar, lecturer in Church History and lay preacher. It is also written by a Taoist who discovered Taoism and Taoist Tai Chi (or they discovered me) over 35 years ago and who has been practising and playing both ever since.[6] It is also written by a student of culture and literature for over 50 years. It stands on the shoulders of those giants who respect and value both spirituality and materiality, such as Walter Benjamin and Henry David Thoreau (perhaps unlikely comrades) and elders past and present of Australian Aboriginal cultures. It is not a book of biblical studies and does not cover or survey the vast literature in the field, but it is in part a book of studies of selected books or aspects of the Bible and their spin-offs in other books, including biblical studies and literary texts.

Environmental Humanities and Theologies is aimed primarily at those who care and are concerned about the natural and cultural environments, but have little or no knowledge of how theological concepts, themes and issues have informed and problematized discussion of those environments and issues around them. It is also aimed mainly at those concerned about the natural and cultural environments who are looking for ways of thinking, being and living with the earth that respect and value both spirituality and materiality. It is a new 'bible' (book of books, the literal meaning of the word 'bible'), a new old and new testament for the symbiocene. It does not seek, and could not hope, to replace the old Bible of the old old testament and old new testament. It has no pretensions to being a sacred text, but aims to nurture sacrality.

Environmental Humanities and Theologies engages with, discusses and critiques selected theological themes, issues and concerns that have infused and confused thinking and writing about environments, and that inform being and living with the earth. It represents the tip of the iceberg of the topics discussed and does not pretend to be exhaustive. It expresses my personal, if not idiosyncratic, interests and concerns with gaps and absences (again, like the biblical writers). It generates what Raymond Williams, the founder of ecocriticism and ecocultural studies, called 'resources (for a journey) of hope'[7] for living with the earth (as land, as planet with air, fire and water) in mutuality and in bio- and psycho-symbiosis, with its plants and animals, lands and waters, weather and climates, people and places. After all, it is our only home.

Notes

1 For further discussion of the philosophical genealogy of the four elements, particularly in relation to Aristotle, see Giblett (1996, pp. 156–160).
2 I discuss Leopold's conservation aesthetics in Giblett (2011, pp. 72–73) and wilderness in Giblett (2011, chapter 5), including Thoreau's 'tonic of wildness' as 'the bog in our brains and bowels', as well as critiquing his racist views of the first American peoples.
3 See Giblett (2011, especially figure 2, pp. 32–34). This book argues for a rapprochement between the two paradigms so that they are brought into complementarity (yin + yang = tai chi, 'the grand ultimate').
4 I am grateful to Glenn Albrecht for the concept of the symbiocene (see Giblett 2016, pp. 12 and 251, n.1). I make the argument and call for living bio- and psycho-symbiotic livelihoods in bioregions of the living earth in Giblett (2011, especially chapter 12).
5 For ecocultural studies (the study of ecoculture, the earthly home-habitat culture, or abode culture to use Harrison's translation of *oikos* (Harrison, 1992, p. 200)), see Giblett (2012). I first proposed psychoanalytic ecology in the 1990s (see Giblett, 1996, pp. 90–91) and developed it in the 2000s (see Giblett, 2009, pp. 2–3 and 131–133). Postmodern and political (socialist and feminist) ecologies are embraced and elaborated in these and other of my previous books.
6 I discuss Taoism, especially the Taoist body, Taoist ecology, the Tao of water and Taoist Tai Chi in Giblett (2008a, chapter 10 and pp. 178–185; 2008b; and 2013, chapter 17). For the teachings and principal text of 'the three religions'' 'school' or tradition of Taoism to which I belong and for the 'style' or form of Taoist Tai Chi and other internal arts of health that I practice, see Fung Loy Kok Institute of Taoism (2008 and 2012) and: https://www.taoist.org
7 For Williams as the founder of both ecocriticism and ecocultural studies see Giblett (2012). For Williams' and others' work on and for hope see Giblett (2009, chapter 9).

References

Bachelard, G. 1983. *Water and Dreams: An Essay on the Imagination of Matter*. E. Farrell, trans. Dallas: Dallas Institute of Humanities and Culture.

Barnhill, K. 2017. *The Girl Who Drank the Moon*. London: Piccadilly.

Beowulf. 1999. S. Heaney, trans. London: Faber and Faber.

Conradie, E., Bergmann, S., Deane-Drummond, C. and Edwards, D. 2014. Discourse on Christian Faith and the Earth. *In:* E. Conradie, S. Bergmann, C. Deane-Drummond and D. Edwards, eds. *Christian Faith and the Earth: Current Paths and Emerging Horizons in Ecotheology*. London: Bloomsbury T and T Clark, pp. 1–10.

Douglas, M. 1966. *Purity and Danger: An Analysis of the Concepts of Pollution and Taboo*. London: Routledge.

Fung Loy Kok Institute of Taoism. 2008. *A Path of Dual Cultivation: Teachings of the Fung Loy Kok Institute of Taoism*. Toronto: Fung Loy Kok Institute of Taoism.

Fung Loy Kok Institute of Taoism. 2012. *Ten Thousand Buddhas Sutra*. Fung Loy Kok Institute of Taoism, trans. Toronto: Fung Loy Kok Institute of Taoism.

Giblett, R. 1996. *Postmodern Wetlands: Culture, History, Ecology*. Edinburgh: Edinburgh University Press.

Giblett, R. 2007. Black and White Water. *In:* E. Potter, A. Mackinnon, S. McKenzie and J. McKay, eds., *Fresh Water: New Perspectives on Water in Australia*, Carlton: Melbourne University Press, pp. 31–43.

Giblett, R. 2008a. *The Body of Nature and Culture*. Basingstoke: Palgrave Macmillan.

Giblett, R. 2008b. *Health Recovery: The* Taoist Tai Chi™ *Way*. London: Shepheard Walwyn.

Giblett, R. 2009. *Landscapes of Culture and Nature*. Basingstoke: Palgrave Macmillan.

Giblett, R. 2011. *People and Places of Nature and Culture*. Bristol: Intellect Books.

Giblett, R. 2012. Nature is Ordinary too: Raymond Williams as the Founder of Ecocultural Studies, *Cultural Studies*, 26(6), pp. 922–933.

Giblett, R. 2013. *Black Swan Lake: Life of a Wetland*. Bristol: Intellect Books.

Giblett, R. 2016. *Cities and Wetlands: The Return of the Repressed in Nature and Culture*. London: Bloomsbury.

Giblett, R. 2018a. *The Dragon and Saint George: A Fairy Tale Novella*. Cambridge: Pegasus.

Giblett, R. 2018b. *Tales of Two Dragons*. London: Austin Macauley.

Habel, N. 2000. Geophany: The Earth Story in *Genesis* 1. *In*: N. Habel and S. Wurst, eds., *The Earth Story in Genesis*, Sheffield: Sheffield Academic Press, pp. 34–48.

Habel, N. 2008. Introducing Ecological Hermeneutics. *In*: N. Habel and P. Trudinger, eds., *Exploring Ecological Hermeneutics*, Atlanta: Society of Biblical Literature, pp. 1–8.

Habel, N. 2011. *The Birth, the Curse and the Greening of the Earth: An Ecological Reading of Genesis 1–11*. Sheffield: Sheffield Phoenix Press.

Harrison, R. 1992. *Forests: The Shadow of Civilization*. Chicago: University of Chicago Press.

Hiebert, T. 2008. Air, the First Sacred Thing: The Conception of רוּחַ in the Hebrew Scriptures. *In*: N. Habel and P. Trudinger, eds., *Exploring Ecological Hermeneutics*, Atlanta: Society of Biblical Literature, pp. 9–19.

Lloyd, G. and Sivin, N. 2002. *The Way and the Word: Science and Medicine in Early China and Greece*. New Haven: Yale University Press.

Tolkien, J. R. R. 2014. *Beowulf: A Translation and Commentary*. C. Tolkien, ed. London: HarperCollins.

Westermann, C. 1984. *Genesis 1–11: A Commentary*. J. Scullion, trans. London: SPCK.

2 Theology of wetlands and marsh monsters

From the Old English *Beowulf* to J. R. R. Tolkien's work, whether as the writer of *The Lord of the Rings* or as critic and scholar of Old and Middle English literature, wetlands and marsh monsters get bad press. *Beowulf* interpolates several verses into the biblical book of *Genesis* that explain the propagation of monsters and that justify regarding wetlands as evil, pagan places. Tolkien's view of marshes and marsh monsters developed partly out of the Old and Middle English and Christian literary precursors he is drawing on in *Beowulf* and *Sir Gawain and the Green Knight*; partly out of his Christian beliefs about wetlands theologized and moralized as places of evil and monsters, a view forged in concert with his critical and scholarly work on *Beowulf* and *Sir Gawain and the Green Knight* and expressed in his portrayal of the Dead Marshes in *The Lord of the Rings;* and partly out of his personal experience of the wet wasteland of mud and slime on the western front in World War I, the prototype for the Dead Marshes in *The Lord of the Rings*.

In this chapter I read and critique the portrayal of wetlands in Tolkien's work and argue that his Christian reading of them did not embrace *Genesis* 1: 1–2 in which God created the world as wetland, nor *Genesis* 2: 6 in which water came up from the aquifers and watered the surface of the ground. Rather, in line with *Beowulf*, he emphasises the wetland and its associated aquifers as a place created after the Fall and, as a result, from which monsters are born and to which evil monsters are condemned. I suggest that this pejorative Christian view of wetlands is largely responsible for the destruction of wetlands in the West for the past millennium.

Tolkien's portrayal of wetlands in these works presents a mode of thinking about the wetland world that is conservatively in line with dominant cultural discourses and is Christian in its genesis, but is out of line with modern scientific, pre-modern sacral and postmodern counter-cultural views of wetlands — perhaps unlikely bedfellows. Wetlands are vital for life on earth, both human and non-human life. The leading intergovernmental agency on wetlands states that:

> they are among the world's most productive environments; cradles of biological diversity that provide the water and productivity upon which countless species of plants and animals depend for survival. Wetlands are indispensable for the countless benefits or 'ecosystem services' that they provide humanity, ranging from freshwater supply, food and building materials, and biodiversity,

> to flood control, groundwater recharge, and climate change mitigation. Yet study after study demonstrates that wetland area[s] and [their] quality continue to decline in most regions of the world. As a result, the ecosystem services that wetlands provide to people are compromised.
>
> (Ramsar Convention Bureau, online)

Yet more than the mere providers of 'ecosystem services', wetlands are habitats for plants and animals, and homes for people. Of course, it is churlish to critique the mistakes of the past from the privileged vantage point of the present. Yet rather than merely bemoan the acts of the past and the facts of history, the point is that some of the mistakes of the past are perpetuated into the present with the continued destruction of wetlands supported by the denigratory views of wetlands in the works mentioned.

By contrast with Tolkien's denigratory view of wetlands, I conclude this chapter by presenting the recent view of Paul Kingsnorth in *The Wake* in which the wetland is a place of burgeoning and decomposing life, of both life-giving and death-dealing water, of sacrality and resistance to hegemony. Diana Gabaldon's *An Echo in the Bone* is another recent instance of a similar view of wetlands, with her account of 'The Great Dismal' Swamp (Gabaldon, 2009, pp. 369–374).

Beowulf's monstrous fens

Tolkien's pejorative Christian view of wetlands was forged in concert with his reading and translation of, and in his remarks and commentary on, *Beowulf*, described by Tolkien (2006, pp. 30–31) himself as 'the most successful Old English poem', more specifically as 'an heroic-elegiac poem', and so not an epic poem, though it has the obligatory descent of the hero in epic poems into the underworld, often a wetland as in this poem. In *Beowulf* Hrothgar describes Grendel the monster and his mother as two:

> huge wanderers of the marches guarding the moors, alien spirits [. . .] They dwell in a land unknown, wolf-haunted slopes, wind-swept headlands, perilous marsh-paths, where the mountain stream goes down under the mists of the cliff, – a flood under the earth. It is not far hence, in miles, that the lake stands over which hang groves covered with frost: the wood, firm-rooted, over-shadows the water. There may be seen each night a fearful wonder, – fire on the flood! [. . .] That is no pleasant spot. Thence rises up the surging waters darkly to the clouds, when the wind stirs up baleful storms, until the air grows misty, the heavens weep.
>
> (*Beowulf*. . . , 1950, pp. 89–90)

Heaney (*Beowulf*, 1999, p. 45) prefers 'uncanny' to 'a fearful wonder', quite aptly in my view as the wetland is the uncanny place par excellence (as I have argued elsewhere; see Giblett,1996, chapter 2). It is 'not far hence in miles' because it is closer than you think in mind. This is a typical trope for the horror of swamps.

Hrothgar's description of this wetlandscape can be compared with a similar, later medieval one in *Sir Gawain and the Green Knight* translated by Tolkien as 'the heavens are lifted high, but under them evilly/ mist hangs on the moor, melts on the mountain' (*Sir Gawain. . .* , 1975, p. 77). The wetland in both texts not only mixes the elements of water and earth, and air and water when the atmosphere is misty in the temperate zone (and dripping wet in the tropics), but also mixes fire and water when marsh gases are ignited in the temperate zone. The traditional dragon breathing fire is arguably just a marsh (monster) 'breathing' ignited marsh gases. Fire is also associated with hell, with the burning fires of brimstone, so the wetland in patriarchal Western culture is a kind of watery hell.

Both Grendel and his mother are alien spirits because they are 'Cain's kindred' (*Beowulf. . .* , 1950, p. 25) condemned by God to the wetland hell. The worst of the two monsters, mother and son, is the son, 'that grim spirit [. . .] called Grendel, the renowned traverser of the marches, who held the moors, the fen and fastness; unblessed creature, he dwelt for a while in the lair of monsters, after the Creator had condemned them'. Grendel, however, has not only been condemned but also been made into the personification, or more precisely 'monstrification', of damnation. He is described as 'the grim and greedy creature of damnation, fierce and furious', and as 'the demon, the dark death-shadow' who 'in the endless night [. . .] held the misty moors' (*Beowulf. . .* , 1950, pp. 26 and 28). Grendel is a demonic agent, or 'angel', of death and an orally sadistic monster who lives in the wetland realm of darkness.

Grendel's mother is not much better as she is described as 'monstrous among womenkind [. . .] who must needs inhabit the dread waters, chilling stream' (*Beowulf. . .* ,1950, p. 84). Grendel's mother lives in the dreadful aquifer, the flood beneath the earth, which is even worse than the flood on or over the earth. Grendel's mother is an instance of what Barbara Creed (1986, pp. 44–70) calls 'the monstrous-feminine', or more precisely the monstrous-maternal, a patriarchal figuring of the mother as monster that denies and represses the gylanic Great Goddess or Mother of the sacral swamps and marshes (see Giblett, 1996; 2011). In patriarchal Western culture the pejoratively coded monstrous maternal marsh and maternal marsh monster by no means began with *Beowulf*, nor ended there (see Giblett, 2008, chapter 5). They go back to classical Greek times with, for instance, the Herculean labour of killing the monstrous Stymphalian birds that live in a marsh. Grendel and his mother are both specifically marsh monsters and a part of a long genealogy in patriarchal Western culture.

Both are also orally sadistic marsh monsters with Grendel variously described as 'the grim and greedy creature of damnation' (*Beowulf. . .* , 1950, p. 26), as 'greedily loping' (*Beowulf*, 1999, 24) and as 'the monster' who 'seized a sleeping warrior' and 'rent him greedily' (*Beowulf. . .* , 1950, p. 58). Grendel is moreover 'a devouring murderer' who 'consumed the beloved man's whole body' (126). Heaney translates this line in part as 'Grendel's maw' (*Beowulf*, 1999, p. 66). Maw is a pejorative term for mouth often used in relation to orally sadistic monsters, such as alligators, crocodiles and sharks. Grendel's mother is also described as 'fiercely ravenous' (*Beowulf. . .* , 1950, p. 96), translated by Heaney as 'gluttonous' (*Beowulf*, 1999, 49) and by Clark Hall as 'greedy and grim' (*Beowulf*. . . , 1950, p. 96). Orally sadistic monsters, I have argued elsewhere, are projections, displacements and disavowing devices for the greed and gluttony meted out to the earth in mining, pastoralism and wetlands dredging and draining, especially as carried out later by orally sadistic monstrous machines that consume the earth (see Giblett, 1996; 2011).

In *Beowulf* Cain, the biblical son of Adam and Eve, and brother and murderer of Abel, is regarded as the progenitor from whom 'all evil broods were born, ogres and elves [Tolkien (2014, pp. 16 and 162) has 'goblins' as elves are good in his view] and evil spirits – the giants also, who long time fought with God' (*Beowulf*. . . , 1950, p. 25; in his lecture on *Beowulf* Tolkien (2006, p. 26) states that Cain is 'the ancestor of the giants'). Tolkien (2014, pp. 161–163) in his published commentary on *Boewulf* claimed that this specifically Christian passage (but one that has no biblical warrant or foundation) is an addition or insertion by the Beowulf poet into the older material he is assimilating for this specific composition. Later in *Beowulf* (1950, p. 85) the poet says that from Cain 'were born numbers of fateful spirits, of whom Grendel was one, a hateful outcast foe [. . .] the monster'. Grendel was descended from Cain by way of Grendel's mother. The *Beowulf* poet effectively interpolates a concoction in *Genesis* chapter 4 between verses 16 and 17 and after the story of Cain's murder of Abel:

> 16 Then Cain went away from the presence of the Lord, and settled in the land of Nod, east of Eden.
>
> 16a Thence all evil broods were born, ogres and goblins and evil spirits – the giants also, who long time fought with God.
>
> 16b From Cain were born numbers of fateful spirits, of whom Grendel was one, a hateful outcast foe, the monster.
>
> 16c That grim spirit called Grendel, the renowned traverser of the marches, held the moors, the fen and fastness; unblessed creature, he dwelt for a while in the lair of monsters, after the Creator had condemned them.
>
> 17 Cain knew his wife, and she conceived and bore Enoch; and he built a city, and named it Enoch after his son Enoch.

Cain gave birth to monsters in the land east of Eden, the land outside the Garden of Eden. Monsters were thus not born in Eden, the sinless pre-lapsarian garden of God, but outside it in the wetlands before the city both temporally and spatially. Cain also gave birth to the city (the post-lapsarian city of sin, the monstrous city) after he gave birth to monsters. The city came after God had completed His creation, including the Garden of Eden, after the Fall and then, after Cain murdered Abel, God cursed Cain and Cain created the city.

Cain is figured as giving birth like Zeus to monsters in perhaps the first recorded instance in English literature of a bachelor birth, though unlike Zeus who gave birth to Pallas Athene out of his head, there is no mention as to which part of his body Cain gave birth to monsters from. Just as the birth of Pallas Athene from the head of Zeus after he had ingested her mother Metis is a Greek patriarchal myth of origins that denies or represses and appropriates the role of the human mother in giving birth to children, so Cain's giving birth to monsters is a Christian patriarchal myth of origins that denies or represses and appropriates the role of the great mother of the marshes in giving birth to wetlands and its creatures. Although there is certainly no biblical warrant or basis for regarding monsters as Cain's kindred and him as their progenitor, Tolkien (2006, p. 19) argued in his lecture on *Beowulf* that these references are vital for the poem or, in his terms, that in them 'the key to the fusion-point of imagination that produced this poem lies'.

The creator certainly condemned Cain for murdering his brother (*Genesis* 4: 10–16). He also condemned the serpent to be cursed among all animals and wild creatures for tempting Eve to eat of the tree of knowledge of good and evil (*Genesis* 3: 14; 2: 17). The serpent was also certainly created by the Lord God, but was 'more crafty than any other wild animal that the Lord God had made' (*Genesis* 3: 1), hence his or her or its predilection to tempt Eve with his/her/its wiles. There is no biblical reference in *Genesis*, however, to other creatures, such as monsters condemned by Him, and certainly not condemned to marshes and swamps, though there is biblical reference to other watery monsters, such as Leviathan in *Job* and the *Psalms* who may be a crocodile, the king of the jungle swamp (as we will see in chapter four below).

Certainly in *Genesis* there were giants or nephilim 'on the earth in those days – and also afterwards [. . .] who were heroes that were of old, warriors of old' (*Genesis* 6: 4). These giants are like Beowulf, the hero of *Beowulf*, though he does not fight with God (unlike the biblical giants) but with the monsters Grendel and his mother. Tolkien (1950, p. xi) remarks that Beowulf has 'an addition of power, beyond the natural', in other words, supernatural; he is a superhero. By contrast, in Heaney's translation Hrothgar describes Grendel as 'an unnatural birth' (*Beowulf*, 1999, p. 45). Tolkien (1950, p. xi) goes on to remark that 'the superhuman thirtyfold strength possessed by Beowulf (in this Christian poem [. . .] is his special gift from God)'. By contrast, the two villainous monsters condemned by God are imbued with deadly demonic power. Grendel is described as 'the demon, the dark death-shadow' (*Beowulf*. . . , 1950, p. 28; *Beowulf*, 1999, p. 7). Whereas Clark Hall calls them 'evil spirits' (*Beowulf*. . . , 1950, p. 25) and Heaney 'evil phantoms' (*Beowulf*, 1999, p. 6), Tolkien (2104, p. 16) in his translation of *Beowulf* calls them 'haunting shapes of hell'. In his accompanying commentary Tolkien traces the etymology of the Old English words back to their associations with 'dead body' and 'the "undead"' (Tolkien 2014, p. 163). Whereas Clark Hall calls them 'mysterious creatures of hell' (*Beowulf*. . . , 1950, p. 28) and Heaney (*Beowulf*, 1999, p. 7) 'reavers from hell', Tolkien (2014, p. 167) has 'sorcerers of hell' and comments on hell as meaning 'the underworld, Hades, the Realm of the Dead [. . .], the darkness'. The swamp as hell and hell as the swamp as the unhomely home for monsters by no means begins with *Beowulf*, nor ends with it (as I have shown elsewhere; Giblett, 1996).

While the story of the serpent, the temptation of Eve and the Fall, and the story of Cain and Abel are all well known, how or where the Christian basis for the story of Grendel and his mother arose or was developed or concocted out of the story of Cain, or even as monsters given birth by Cain, is another question. The reason for it lies, I think, in the shift from the sacrality of wetlands to Christianity as a religion of drylands. One way in which Christianity asserted its dominance in England over the sacrality of swamps and marshes was figuring and condemning wetlands as hell, as a place of monsters, the basis for later draining and 'reclaiming' them. Theologically *Beowulf* is highly suspect, but evangelistically it was very effective in asserting the dominance of Christianity and ecologically very destructive of wetlands in asserting the dominance of dryland agriculture.

The representative landscape of modern European Christianity is a drained wetland, first made possible by wind-powered pumps and later, in the modern industrial age, by steam-powered dredgers. The drained wetland represents the triumph of Christian beliefs and morality over the people of the Fens and other wetlands who valued wetlands as the place of sacrality par excellence (as I have shown previously; see Giblett, 1996).

Wetlands were subjected to two double whammies in and by patriarchal society and culture. First, wetlands were inimical to the city. The hard and heavy materials of European-style cities cannot be supported in and by the soft and moist soils of wetlands. The establishment of cities in wetlands, such as London, Paris and Venice, involved the draining and filling, or canalising and reclamation of wetlands as the foundational event for establishing European cities, later their settler diasporas, and for the age of the cities that is still ongoing. The city triumphs over the swamp, the metropolis over the marsh. The repressed wetland returns, however, in floods, as with New Orleans in the aftermath of Hurricane Katrina (see Giblett, 2009, chapter 10; 2016, chapter 10). Secondly, wetlands were denigrated and demonized by Christianity in the foundational event for the establishment of Christian hegemony over sacrality in both Europe and its colonies. Thirdly, wetlands were drained and filled by wind-powered pumps and later steam-powered dredgers in the foundational event of the modern age, with the development of modern industrial agriculture and cities, followed later by the expansion of suburbia in megapolises into wetlands. To add insult to injury, the slums and the dark underside of modern cities were also figured as wetlands. The repressed wetland returns in figures for slums and the dark underside of the modern city. Fourthly, and as we will see below, modern industrial technology, agriculture, cities and war created wet wastelands in the foundational event of the hypermodern age. To add further insult to injury, the wet wastelands of World War I were also figured as wetlands, as I will show shortly with Tolkien's 'Dead Marshes' in *The Lord of the Rings*.

Beowulf plays a perhaps unlikely, but vital, role in this history. Regarding the Christian (and hybrid Germanic) provenance and parentage of *Beowulf*, Wrenn leaves us in no doubt when he argues in his introduction to the Clark Hall translation of *Beowulf* that the writer is:

> a Christian poet addressing a Christian audience, yet sharing with that audience a deep interest in history, legend and tradition, and presenting the inherited Germanic heroic ideals in an inherited poetical diction, [who] would not go out of his way to include anything *specifically* Christian: yet for him as for his listeners, the traditional ideals and ways of life must inevitably be 'coloured' by the new Christian Latin culture.
>
> (*Beowulf*. . . , 1950, p. 9)

Yet the poet *did* go out of his away to include the specifically European Christian (but dubiously biblical) account of the birth and condemnation of monsters which *was* coloured with a distinctly dark Germanic hue with its account of the birth of monsters and the monstrification and demonizaton of wetlands.

Along similar lines to Wrenn, Tolkien (2006, p. 22) stated in his lecture on *Beowulf* that 'if the specifically Christian was suppressed [and refers in a footnote to 'the only definitely Scriptural references' to the story of Cain and Abel (Tolkien, 2006, p. 45], so also were the old gods'. This strikes me as disingenuous for despite their dubious biblical basis the specifically Christian references in *Beowulf* to monsters as Cain's kindred and to him as progenitor of monsters do work precisely and explicitly to suppress 'the old gods'. And goddesses. Tolkien (2006, p. 26) later states that in *Beowulf* 'we have [. . .] an historical poem about the pagan past'. Not only are its claims to historical veracity about 'the pagan past' shaky, as Tolkien acknowledges and as one would expect in an historical poem, but also its take on 'the pagan past' is problematic given that the old gods of monsters were not stamped out by the agents of the new God and so 'the pagan past', or sacrality, is still present. By contrast, Kingsnorth's *The Wake* is a historical novel about the sacral present set in the past (which I will show later).

In his commentary on *Beowulf* accompanying his translation Tolkien (2014, p. 159) comments selectively on these passages and on the provenance of the monsters. He does not comment on their uncanny wetland (un)home. The setting for the poem is the mere background or backdrop against which the Christian story and heroic actions of Beowulf are set. It is not the context for the story, nor is it the environment in which the story takes places, let alone being an agent or actor in the plot (as is his portrayal of the Dead Marshes in *The Lord of the Rings* as we will see shortly), nor are the context and environment agents or actors for Tolkien when he comes to discuss the second part of Beowulf in which the eponymous hero and a dragon do battle (as we will see in the following chapter).

Yet the association between the monsters and their marshland (un)home is vital for the story (as it is for Frodo in *The Lord of the Rings*), especially as Beowulf descends like the hero in the epics of Homer, Virgil and Dante into the underworld of the marsh to slay the monsters. Tolkien (2014, pp. 16 and 158) comments at length on the anachronistic Miltonic description of Grendel as 'a fiend of hell', glossed by Clark Hall as 'like all other monsters, he was of the brood of Cain, and dwelt, an outcast spirit, among the moors and fens' (*Beowulf*. . . , 1950, p. 25). In a much more secular, but also theological key, Heaney (*Beowulf*, 1999, p. 50) in his translation of *Beowulf* translates a later line about Grendel's mother as 'that swamp-thing from hell'. Precisely. I wonder if Heaney knew about the *Swamp Thing* comic series published by DC comics. The swamp as hell (and vice versa) is a persistent trope.

Tolkien (2014, pp. 159 and 161) wonders whether 'a fiend of hell' is possibly 'a kind of half-theological notion' from 'an unwritten chapter (as it were) of the Old Testament'. Yet despite its dubious theological basis, at best half-theological and half-mythological, at worst an anti-sacral Christian concoction, this unwritten chapter of the Old Testament *had* to be written so European Christianity could write the epic story of its heroic conquest of pre-Christian sacral marshes and their monsters. This chapter *was* written and the chapter is called *Beowulf.* It became a canonic text, not only of English literature but also of English Christianity, with its new creation story about marshes and monsters interpolated in *Genesis* 4: 16 after Cain's punishment by God and before *Genesis* 4: 17 and Cain's creation of the city.

Cain is the father of monsters, the father of both monsters in the marsh and the monstrous city in the marsh. The city is like a son; the city drains the marsh before it and eradicates the monster while taking up and becoming monstrous itself, true to its conception. The monstrous city destroyed the monstrous marsh, the dryland city dredged and drained the wetland. Perhaps this is a case of history repeating itself, of the sin of the father of fratricide, being repeated and revisited on the next generation with one of his progeny – the city – killing another—the marsh monster and monstrous marsh.

Cities built on or by wetlands are not only the work of fallen 'man' in creating something that God did not create, but also of fallen 'man' who rises up and wrests from God the divine function of dividing land from water (*Genesis* 1: 6), and becomes as God. The builders of the ancient city of Babel became as God by building a city that rose to the heavens; the builders of the ancient, medieval, Renaissance and modern cities set in wetlands became as God by building a city that divided land from water immersed in the earth. However, they could not become as God by creating wetlands; in fact, they destroyed wetlands in and by creating the city. Only recently has 'man' become as God by creating wetlands. 'Man' thus arrogated to himself the power to both destroy and create wetlands and thereby became more than God, a super God.

Beowulf as the half-written new chapter in Old English of the Old Testament would be posed in the context of the whole book of *Genesis* against *Genesis* 1: 1–2 in which God created the world as wetland, though these verses had never been read to my knowledge in this way until I first did so over twenty years ago (Giblett, 1996, pp. 142–143). In *Postmodern wetlands* and in the previous chapter I argued that the wetland harks back to the time when the entire earth was wet land. I also argued that *Genesis* 1: 1–2 needs to be rewritten or retranslated in pre-modern sacral and postmodern counter-cultural and ecological terms that recognizes the importance, even primacy, of wetlands: 'in the beginning God created the heavens and the earth, and the earth was wetlands.' The creation of the world as wetland was God's first and best work. When God divided the land from the water in *Genesis* 1: 9–10 it was not so much that He forgot or forsook wetlands than that He knew that the land and the water and the creatures that He later created to live in or on them would not survive and would not be able to reproduce without wetlands. Subsequent days of creation do not supersede and 'forget' about their predecessors but build and rely upon them. Indeed, before God sent rain water came up from the aquifers and watered the surface of the ground. In the Judeo-Christian creation story the days have been seen in terms of a progressive timeline heading towards the ultimate goal of humanity, and not as an evolutionary process.

Without the first day of creation, the creation of the earth as wetland, there is no sixth day, no creation of humankind. Without the recreation of wetlands on the first day, and every day for that matter, there is no recreation of human life, there is no human life full stop, on the sixth or any other day. Rather than being forgotten by God, the wetland is the first day of creation, God's first and best work, without which no other day could ever begin, could ever dawn. Rather than being hell, the wetland is the closest thing to heaven because it was created next, right after heaven. When one goes to the wetland one goes back to the first day of creation when it springs fresh and new from the hand and word of God, when the spirit broods over the earth as wetland like a hovering marsh harrier, not like a malevolent ghost or swamp monster.

Canadian Wetlands: Places and People begins with a rewriting of *Genesis* 1: 1–10 and 4: 16a-c:

> In the beginning there was the sky above, and the earth and water below. In the middle was the wetland. The earth and the water were without form and were chaotic. Darkness and light moved over the face and body of the earth and water. Earth and water were wet land. This was the first act of creation, the first coming into being, the first and best work of creation. The world was wetland, and the wetland was womb from which all later life sprung. Including human life made, not out of dust, into and for a desert people, but out of mud into and for marsh people. The wetland is the womb, and the womb is a wet land; both are a slimy swamp of embryonic life. The wetland is not only the womb but also the tomb into which all life died and from which new life was, and is, born. This was the world before the fall when the world was good, before the world fell into the knowledge of good and evil. This was world before the swamp and the marsh became the home alone of darkness, disease, death and to grotesque monsters lurking in the uncanny depths of their murky waters evoking horror and fascination in any who should be so (un)fortunate to stumble into them.
>
> (Giblett, 2014, p. 3)

Tolkien's Dead Marshes

Tolkien certainly buys into and reproduces the doctrine straight out of *Beowulf* that wetlands were the creation of evil after the Fall and that they were not created by God before the Fall. In *The Silmarillion*, the back-story and prequel to

The Lord of the Rings, as a consequence of the 'evil doings' of Melkor, 'green things fell sick and rotted, and rivers were choked with weeds and slime, and *fens were made*, rank and poisonous, the breeding place of flies' (Tolkien, 1977, p. 36; my emphasis). Fens were a consequence of evil. They did not exist prior to evil. This is in line with *Beowulf* and its new chapter in Old English of the Old Testament.

Tolkien's take on wetlands achieves its ultimate expression and pejorative representation in the 'Dead Marshes' in *The Two Towers*, the second volume of *The Lord of the Rings* trilogy, in a chapter called 'The Passage of the Marshes'. Here Tolkien has his narrator describe the entry of Frodo, Sam and Gollum into these marshes via a stream where:

> The gully became ever shallower and the slope of its floor more gradual. Its bottom was less stony and more earthy, and slowly its sides dwindled into mere banks. It began to wind and wander.
>
> (Tolkien, 2002, p. 630)

This description of the transition from stream to marsh, of stream morphing into marsh, of a stream petering out in a marsh enacts a number of conventional geomorphological preferences, or prejudices: deep over shallow; incline over gradual slope; stony or hard bottom over soft or earthy bottom; defined sides over dwindling banks; and a straight course over a winding and wandering way.

It only gets worse:

> That night drew to its end, but clouds were now over the moon and star, and they knew of the coming day only by the slow spreading of the thin grey light. In a chill hour they came to the end of the water-course. The banks became moss-grown mounds. Over the last shelf of rotting stone the stream gurgled and fell down into a brown bog and was lost. Dry reeds hissed and rattled though they could feel no wind.
>
> (Tolkien, 2002, p. 630)

The Dead Marshes are not only a liminal place between land and water but also a twilit crepuscular place between night and day, dark and light. The coursing water ends and stops flowing. The dwindling banks morph into moss-grown mounds. The stream stops abruptly and is lost in a brown bog. The green world is lost in brown wetland, or wasteland for Tolkien. Despite being a wetland, dry reeds hiss like snakes or serpents.

The Marshes continue to get worse:

> On either side and in front wide fens and mires now lay, stretching away [. . .] into dim half-light. Mists curled and smoked from dark and noisome pools. The reek of them hung stifling in the still air.
>
> (Tolkien, 2002, p. 630)

The initial sensory encounter entering the Dead Marshes is via the distancing sense of sight, whereas this more immediate experience is via the up-close sense of smell. Typically the marsh is a place of miasma and effluvia.

And so it goes on:

> It was already day, a windless and sullen morning, and the marsh-reeks lay in heavy banks. No sun pierced the low clouded sky [. . .] They were soon lost in a shadowy silent world, cut off from all view of the lands about [. . .] What had looked liked one vast fen was really an endless network of pools, and soft mires, and winding half-strangled watercourses.
>
> (Tolkien, 2002, p. 632)

Explorers often complained about getting lost in wetlands because they obstructed the view.

To reach the nadir:

> It was dreary and wearisome. Cold clammy water still held sway in the forsaken country. The only green was the scum of livid weed on the dark greasy surface of the sullen water. Dead grasses and rotting weeds loomed up in the mists like ragged shadows of long-forgotten summers.
>
> (Tolkien, 2002, p. 632)

The marsh becomes 'it', a thing, a swamp thing, and ceases to be a living being. It is also melancholic. It not only assaults and offends the sense of smell but also affects adversely the psychology of the marsh traveller. It is figured as a place of death, whereas it is a place of both life and death, life decomposing, dying and regenerating into new life. I am reminded of the souls of the sullen stuck in slime in Dante's *Inferno.*

Yet there is always a beacon of Christian light in this benighted place:

> As the day wore on the light increased a little, and the mists lifted, growing thinner and more transparent. Far above the rot and vapours of the world the Sun was riding high and golden now in a serene country with floors of dazzling foam, but only a passing ghost of her could they see below, bleared, pale, giving now colour and no warmth [. . .] There was a deep silence, only scraped on its surfaces by the faint quiver of empty seed-plumes, and broken grass-blades trembling in small air-movements that they could not feel.
>
> (Tolkien, 2002, p. 632)

The narrator takes a privileged, God-like position above the quaking zone of the wetland scene to describe the sun shining on all below her. The sun is feminized in line with the female Maia Arien who drives the chariot of the sun in *The Silmarillion*, as opposed to the male Maia Tillion who drives the chariot of the moon. This 'gendering' of the sun and moon is already there in the earliest version of Tolkien's mythology in *The Book of Lost Tales* composed in the 1910s–1920s. What is the marsh? How is the marsh gendered? Typically and traditionally it has been feminized.

The marsh is a place not only of melancholy but also of mourning:

> 'Not a bird!' said Sam mournfully. 'No, no birds,' said Gollum [. . .] 'No birds here. There are snakeses, wormses, things in the pools. Lots of things, lots of nasty things. No birds.'
>
> (Tolkien, 2002, p. 632)

The marsh is a place of lower creatures in the great Christian hierarchical chain of being.

The wetland becomes even wetter:

> The fens grew more wet, opening into wide stagnant meres, among which it grew more and more difficult to find the firmer places where feet could tread without sinking into gurgling mud [. . .] Presently it grew altogether dark: the air itself seemed black and heavy to breathe.
>
> (Tolkien, 2002, p. 633)

Garth (2006, p. 46) comments on this and other passages in *The Lord of the Rings* about 'the very air of the nightmarish Dead Marshes' that 'there are grounds to suspect that Tolkien was influenced by his experience [as a soldier in the World War I] of poison gas as he devised a symbolic shape for battlefield trauma, demoralization and despair'. Yet the heavy air and marsh gases of wetlands is not the result of poison gases produced by modern industry and modern warfare, but of decomposing organic matter.

Tolkien does not distinguish (and Garth follows suit) between on the one hand a wetland as a native quaking zone made by ancestral beings and traditional human beings, and on the other hand a wet wasteland as a feral quaking zone made by modern industrial hands, such as in the case of Somme mud of the World War I that he experienced personally. Tolkien (1981, p. 303) wrote in a letter that he did not think:

> either [the first or second world] war [. . .] had any influence upon either the plot or the manner of its [*The Lord of the Rings*] unfolding. Perhaps in landscape. The Dead Marshes and the approaches to the Morannon owe something to Northern France after the Battle of the Somme.

After citing this extract, Garth (2003, p. 310) goes on to relate that for Hugh Cecil in his survey of British fiction of World War I the Dead Marshes is 'a scene of morbid desolation that has become in effect, a shorthand symbol for the trenches' (Garth, 2003, p. 311). The Dead Marshes were not inspired by a living, native wetland, but rather by a dead, feral wet wasteland, a counter-intuitively artificial wetland (see Giblett, 2009, chapter 1). The Dead Marshes then became a shorthand symbol for the scene of morbid desolation of the feral wet wasteland of the mud of the Somme created by modern industrial warfare and its technologies (see Giblett, 2009, chapter 4). Unlike the Somme in World War I, the Dead Marshes were not created by war, but like the Somme they were a burial ground for the war dead.

What could be called a circularity of figurality occurs here in what I call placism: a recently human-made wet wasteland, such as Somme mud, is made to figure a native or ancient wetland, such as the marsh that the Dead Marshes once were, and the pejorative associations of the former are applied to the latter. Both are places of death and the dead, whereas marshes are places of both life and death. The obverse also applies where the pejorative associations of a swamp are applied to the city slum and consequently the swamp comes to be seen in the same pejorative light (or darkness) as the city slum (as we will see in chapter 6). Both are places of death and the dead, whereas marshes are places of both life and death. This pejorative figuring of wetlands becomes mutually reinforcing and culturally naturalized, but its logic is circular and its ideology is placist. Tolkien and other placist writers (as we will see in later chapters) not only ascribe moral qualities to a place undeserving of such opprobrium, but also give it derogatory qualities in the lower psychopathological register.

The history of the Dead Marshes bears out the conclusion that they were created as a burial ground for the dead. The Tolkien Gateway says that:

> In the year S. A. 3434 the host of the Last Alliance of Elves and Men fought the forces of Mordor in the Battle of Dagorlad. During the battle on the plains more than half of the Elves of Lothlórien under the command of King Amdír were driven into the Dead Marshes. After the battle many of the slain were buried outside of the marshy area but over time (in the Third Age) the Marshes had grown and swallowed the graves. In T. A. 1944 King Ondoher of Gondor was caught by a surprise attack by the Wainriders upon the Dagorlad. When the King and his guard were destroyed, many of the confused soldiers of Gondor were driven by the attackers into the Dead Marshes. However, after Eärnil II won the Battle of the Camp, those of the Wainriders who were not slain in the fight were themselves driven into the Dead Marshes and there perished.
>
> (http://tolkiengateway.net/wiki/Dead_Marshes)

Frodo, Gollum and Sam have entered the quaking zone where the earth trembles. They have also entered the dead zone of 'dark water', 'slime', 'sticky ooze' and the marsh lights of burning marsh gases described by Gollum as 'candles of corpses' for this is a place of the dead with 'dead faces in the water' (Tolkien, 2002, p. 633). 'Fire on the flood' has a direct prototype in *Beowulf* (*Beowulf. . .*, 1950, p. 90) for the burning marsh gases in *The Two Towers*. Tolkien survived Somme mud just as Frodo, Sam and Gollum at last:

> came to the end of the black mere, and they crossed it perilously, crawling or hopping from one treacherous island tussock to another. Often they floundered,

> stepping or falling hands-first into waters as noisome as a cesspool, till they were slimed and fouled almost up to their necks and stank in one another's nostrils. It was late in the night when at length they reached firmer ground again [. . .] [Gollum] in some mysterious way, by some blended sense of feel, and smell, and uncanny memory for shapes in the dark, [. . .] seemed to know just he was again, to be sure of his road ahead.
>
> (Tolkien, 2002, pp. 634–635)

The human-made cesspool is made to figure the smell of the natural marsh, albeit in this case a burial ground. In the Dead Marshes Gollum did not know where he was and he was unsure of the road ahead for there was no road. The Dead Marshes are a trackless and roadless waste in which it is easy to get lost. Gollum does not rely on the aesthetic senses of sight and hearing to find his way again but on the uncanny senses of smell and touch to feel and intuit his way.

Kingsnorth's *The Wake*

Published in 2014 like Tolkien's translation of, and commentary on, *Beowulf* and written in what Kingsnorth calls, and devised as, 'shadow English', 'a pseudo-language' that might have been that adumbrates Old and Modern English, *The Wake* is a historical novel set in the aftermath of the Norman invasion of England in 1066, and geographically in and around the Fens. Unlike Tolkien for whom the marsh is the mere backdrop for human action, or the home of horrific monsters, or a horrific place itself, for Buccmaster, the first-person narrator of *The Wake*, 'the fenn' 'was 'what has macd us who we is as folc this place' (Kingsnorth, 2014, p. 33). The shallow waters of the wetland are fascinating and delightful as they are referred to as 'the deorc undeop waters of the fenn' (p. 34). The dark waters of the wetland are not an object, or abject to use Kristeva's more apt term, of horror that hides lurking monsters unlike for the *Beowulf* poet and Tolkien. Buccmaster acknowledges that 'the fenn that can be so blaec and deop and cold on this mergen was a thing of great wundor' (p. 50). Buccmaster has heard the Beowulf story. When he is a child he goes out in his grandfather's boat into the fen and fears that the world under the water is where Grendel and his mother were living and that they 'was cuman for me' (p. 51). The world under the water is the home, not to monsters, but to the old sacrality and folk knowledges. It is also the site of resistance to the Normans as it was for the historical Hereward of the same period usually known as 'the wake', but Kingsnorth seems to be implying or suggesting that Buccmaster is another wake, or the real wake.

The fens are not necessarily a pleasant place but they are subject to commonplace misconceptions. Buccmaster relates 'it is saed that the fenns is all deorc waters and deope mud' and 'that folcs of the fenns' is 'yfel wights of the mere'. The fens are not the place of evil monsters. Buccmaster counters these misconceptions:

> These is the things folcs saes who does not cnawan naht of what they is specan for the fenns is a place of wundor to those who cnawan them. yes there is

> deop meres and waters so blaec that oxen is lost in them and nefer seen and there is muds what strecces for miles by secg and lesch [reeds]. . . there is the blaec fenns where the eorth and the waters in all blaec like the graef.
>
> (Kingsnorth, 2014, p. 123)

Buccmaster acknowledges the unpleasant aspects of the fens, including their black water.

Buccmaster initially seeks shelter from the Normans in the wood but later comes to the realisation that 'i was a man of the water not the wud the fenns was callan me' and that he must go to 'the water lands for there was no other place triewe in this world' (Kingsnorth, 2014, p. 286). When he returns to the fens it is a return to his homeland with familiar black earth, meres and sedges and 'in the heofon was the eald feel of my life and of my folc' (p. 287). He acknowledges that 'i missed the fenns' and that 'it is the triewest place in angland' (p. 287). The fens are the quintessential English homeland, the repository of sacrality. Long may they live.

References

Beowulf and the Finnesburg Fragment. 1950. J. Clark Hall, trans. New edition completely revised with an introduction by C. L. Wrenn and with prefatory remarks by J. R. R. Tolkien. London: George Allen and Unwin.

Beowulf. 1999. S. Heaney, trans. London: Faber and Faber.

Creed, B. 1986. Horror and the Monstrous-Feminine: An Imaginary Abjection. *Screen*, *27*(1), pp. 44–70.

Gabaldon, D. 2009. *An Echo in the Bone: A Novel*. London: Orion.

Garth, J. 2003. *Tolkien and the Great War: The Threshold of Middle-Earth*. New York: Houghton Mifflin.

Garth, J. 2006. Frodo and the Great War. *In*: W. Hammond and C. Scull, eds. *Lord of the Rings, 1954–2004: Scholarship in Honor of Richard E. Blackwelder*, Milwaukee, WI: Marquette University Press, pp. 41–56.

Giblett, R. 1996. *Postmodern Wetlands: Culture, History, Ecology*. Edinburgh: Edinburgh University Press.

Giblett, R. 2008. *The Body of Nature and Culture*. Basingstoke: Palgrave Macmillan.

Giblett, R. 2009. *Landscapes of Culture and Nature*. Basingstoke: Palgrave Macmillan.

Giblett, R. 2011. *People and Places of Nature and Culture*. Bristol: Intellect Books.

Giblett, R. 2014. *Canadian Wetlands: Places and People*. Bristol: Intellect Books.

Giblett, R. 2016. *Cities and Wetlands: The Return of the Repressed in Nature and Culture*. London: Bloomsbury.

Kingsnorth, P. 2014. *The Wake*. London: Unbound.

Ramsar Convention Bureau. The Importance of Wetlands. Accessed online: http://www.ramsar.org/about/the-importance-of-wetlands

Sir Gawain and the Green Knight [. . . .] 1975. J. R. R. Tolkien, *trans*. London: George Allen and Unwin.

Tolkien Gateway, The. Accessed online http://tolkiengateway.net/wiki/Dead_Marshes

Tolkien, J. R. R. 1950. Prefatory Remarks. In: *Beowulf and the Finnesburg Fragment*, J. Clark Hall, trans., new edition completely revised with an introduction by C. Wrenn, London: George Allen and Unwin, pp. ix-xliii.

Tolkien, J. R. R. 1977. *The Silmarillion*. London: George Allen and Unwin.

Tolkien, J. R. R. 1981. *Letters of J. R. R. Tolkien: A Selection*. H. Carpenter, ed. with the assistance of C. Tolkien. London: George Allen and Unwin.

Tolkien, J. R. R. 2002. *The Two Towers: Being the Second Part of the Lord of the Rings*. London: HarperCollins. First published 1954.

Tolkien, J. R. R. 2006. Beowulf: The Monsters and the Critics. Chap. 1 in *The Monsters and the Critics and Other Essays*, C. Tolkien, ed, London: HarperCollins, pp. 5–48. First published 1936.

Tolkien, J. R. R. 2014. *Beowulf: A Translation and Commentary*. C. Tolkien, ed. London: HarperCollins.

3 Theology of dragons and monstrous serpents

The dragon is a legendary figure in many cultures. In English culture and literature the dragon is traditionally a flying, evil, masculine monster who wreaks death and destruction with his fiery breath. From the dragon in *Beowulf* to Tolkien's writings on *Beowulf*, his version of *The Legend of Sigurd and Gudrún* and to his own Smaug in *The Hobbit*, the dragon is a monstrous serpent who lives underground, flies at night or twilight, and embodies and exhales fire and/or venom. Just as wetlands and marsh monsters in *Beowulf* and Tolkien's work get bad press, so do dragons in these writings. The dragon is portrayed as a creature of primeval fiery chaos and as an earth monster who is a terrible hoarder of precious minerals and metals. The dragon in the legend of St George and Satan in Milton's *Paradise Lost* is also a monstrous serpent. The dragon is portrayed as a creature of primeval watery chaos and as a marsh monster who is a horrible creature of slimy swamps. Other dragons appear elsewhere, such as in Malory's *Le Morte d'Arthur* and Bunyan's *Pilgrims Progress* (discussed in a later chapter), with a variety of different features of their anatomy and physiology.

In this chapter I give a psychoanalytic ecological critique of the portrayal of dragons in these texts as evil, orally and anally sadistic monsters onto which are projected and displaced human desires and fears of an oral and excremental nature. I argue that the earth dragons in *Beowulf* and *The Hobbit* are keepers and guardians of minerals and metals in the earth. The earth dragon can be read as an embodiment or personification of the mineral treasures within the earth. The earth dragon guards these treasures with volcanic fire within the earth. Humans steal minerals and metals from the earth/dragon, portray earth dragons as robbers and evil monsters, heap their guilt for doing both on to them as scapegoats and then kill them and waste the earth to expiate their guilt.

Earth dragons on this view are projections of human guilt for robbing the riches of the earth. Humans displace their guilt for greedily and ungratefully stealing and consuming (eating) the mineral treasures or good things of the earth and laying the earth waste (defecating) onto the figure of the dragon who, as in *Beowulf*, ostensibly steals from humans, and humans steal back what he has stolen. The dragon, as in *Beowulf* also, wreaks vengeance with fire on humans for doing so by defecating or spewing volcanic fire or bad things onto them. Humans, as in *Beowulf* too, wreak vengeance on the dragon in return as a way of displacing and projecting onto the dragon their guilt for stealing from him and out of fear that the dragon/earth may wreak vengeance on them for stealing minerals or good things from him/her in the first place.[1] A more environmentally and animal friendly view of earth dragons would see them as manifestations of the precious metals and minerals of the earth to live with and use sacrally and symbiotically, rather than to kill (see Giblett 2018b).

Earth dragons wreaking fiery vengeance on humans for stealing from the earth can also be read as a climate change allegory. More intense bush and forest fires in the age of global warming are the earth's vengeance for greedy mining and carboniferous capitalism. Living more sacrally and symbiotically with dragons as creatures of fire and the earth as combustible means drastically reducing greenhouse gas emissions.

Swamp dragons found in the legend of St George and the dragon and in Milton's *Paradise Lost* function culturally and environmentally in a similar way to earth dragons. Humans steal new life from the fertile swamps/dragon, portray swamp dragons as evil monsters, heap their guilt for doing both on to them as scapegoats and then kill them and drain swamps to expiate their guilt. Swamp dragons on this view are projections of human guilt for living parasitically on the good things of the fertile swamps. A more environmentally and animal friendly view of swamp dragons would see them as manifestations of the fertility of the earth to live with and use sacrally and symbiotically, rather than as objects to kill (see Giblett 2018a and 2018b).

Beowulf's earth dragon

After killing Grendel and his mother, the marsh monsters, and ruling for fifty years, Beowulf is then confronted in *Beowulf* by another dangerous creature and threat to his rule in the form of 'the dragon'. This dragon is a treasure hoarder who is aroused to fiery wrath by some of his treasure being stolen by a slave. The dragon wreaks fiery vengeance on humans by destroying their homes, including Beowulf's great hall. Beowulf is naturally aroused to seek vengeance on the dragon in an 'eye for an eye, tooth for a tooth, tit for tat' payback against the dragon. In the ensuing fight, things go terribly wrong when Beowulf's sword breaks against 'the enameled scales' of 'the slick-skinned dragon', as Seamus Heaney puts it in his colourful translation of *Beowulf* (1999, pp. 72 and 81).

Things go from bad to worse when the dragon clasps Beowulf around his neck with his sharp and venomous fangs. In the gruesome fight to the finish not only does Beowulf's blood gush out of his presumably severed jugular vein, but also the poisonous venom of the dragon suppurates inside him, again in Heaney's vivid translation (*Beowulf*, 1999, p. 85). Beowulf finally succeeds in killing the dragon with his dagger before succumbing to loss of blood, or poison, or both, and dying.

The dragon is a lively player, an active agent and a central character in the story whose qualities are vital to its drama, but Tolkien gives him short shrift and scant regard. In his lecture on *Beowulf*, Tolkien comments that:

> there are [. . .] many heroes but very few good dragons. Beowulf's dragon, if one wishes really to criticize, is not to be blamed for being a dragon, but rather for not being dragon enough, plain pure fairy-story dragon. There are in the poem some vivid touches of the right kind [. . .] in which this dragon is real worm, with a bestial life and thought of his own, but the conception, none the less, approaches *draconitas* rather than *draco*: a personification of malice, greed, destruction (the evil side of heroic life), and of the undiscriminating cruelty of fortune that distinguishes not good or bad (the evil aspect of all life).
>
> (Tolkien, 2006, p. 17)

This is a fairly (Christian?) grim view of the dragon as a personification of the qualities of malice, greed and destruction, as well as of the cruelty of fortune and 'the evil aspect of all life'. A variety of other monstrous animal figures, such as marsh monsters (as we have seen in the previous chapter) and crocodiles (as we will see in the following chapter), have been made to play these roles in 'heroic life'. Tolkien does not mention that the dragon in *Beowulf* is also the manifestation of the mineral treasures and fire within the earth (larval flows, for instance) and so is also the personification of munificence, generosity and creation. This view is akin to the traditional Chinese Taoist one in which the dragon is the manifestation of the complementarity of hard and soft, strength and flexibility, *yin* and *yang* (Ingersoll, 2014, p. 64). In the Taoist arts of health, such as *tai chi*, the dragon represents the spinal column of the body and the mountain ranges of the earth in *feng shui*.[2] The dragon is the backbone of the body of the earth.

The dragon in *Beowulf* is characterised as bad and Tolkien does not have anything good to say about him, or anything at all to say about the 'very few good dragons' and who they are. There are also probably few extant wise dragons.[3] Tolkien's conclusions do not address the rich characterization of the dragon in *Beowulf*. They downplay the dragon's importance for the story and do not acknowledge the dragon's other qualities, such as the personification, or embodiment, or 'en-dragoning', of fire. Tolkien's account of *Beowulf's* dragon, if one wishes really to criticise, is to be blamed for giving the dragon scant regard.

For instance, when it seems Tolkien is about to discuss the dragon, he discusses the dragon-slayer instead:

> As for the dragon: as far as we know anything about these old poets, we know this: the prince of the heroes of the North, supremely memorable [. . .] was a dragon-slayer. And his most renowned deed [. . .] was the slaying of the prince of legendary worms.
>
> (Tolkien, 2006, p. 16)

In other words, the most renowned deed of dragon-slayers was slaying dragons. Dragon-slayers slay dragons in an astonishing tautology! The primary quality and function of the dragon for Tolkien is to be slain by dragon-slayers. The dragon functions for Tolkien primarily as a prop and plot device for the dragon-slayer to slay. The dragon exists to be killed in a display of heroic virtue, strength and dexterity. The dragon in *Beowulf* has few other remarkable qualities that Tolkien notes or discusses in this lecture ostensibly devoted in part at least to 'the monsters' of *Beowulf* as announced in his title.

One quality that Tolkien might have discussed is that the dragon is a worm. Tolkien adopts a standard, cognate translation of Old English 'wyrme' as 'worm' and reiterates the connection on the same page. Elsewhere Tolkien refers to the dragon as 'the serpent' and generally translates the Old English 'wyrme' as 'serpent' in another standard translation, including his own translation first published in 2014 (Tolkien, 2014). What is worm-like about the dragon (living in darkness underground, slithering and sliding along as one of his forms of propulsion (he also flies), ingesting and excreting dead matter, even aerating and fertilizing the earth, etc.) are not on Tolkien's menu for discussion. He seems to avoid them as topics of conversation (Tolkien's comments about the dragon in *Beowulf* were first made in a lecture to the British Academy in 1936. Perhaps the lecture was a dinner speech and Tolkien did not wish to put his listeners off their pudding). Tolkien evinces no curiosity about these and other features of 'the worm'.

Tolkien (2006, p. 16) argues that 'a dragon is no idle fancy. Whatever may be his origins, in fact or invention, the dragon in legend is a potent creation of men's imagination', but he does not discuss how the dragon of *Beowulf* is potently created, what its features are, and so on, despite the *Beowulf* poet going into considerable detail and riffing on a variety of recurring themes, such as that the dragon is a flying monster of the twilight, night, and the sky who breathes out fire from his mouth. Elsewhere Tolkien does note one feature of 'the worm'. In 'The Prefatory Remarks' Tolkien wrote for John Clark Hall's translation of *Beowulf*, he describes the dragon as 'gliding', which could refer to flying, or slithering, or both (Tolkien, 1950, p. 149). Heaney translates 'scríðan' as gliding in a standard translation (*Beowulf*, 1999, p. 81; Heaney, 2007, pp. 172 and 173). Gliding is ambiguously sky-borne flying like a bird or earth-borne slithering like a serpent. Gliding is an appropriate descriptor for the dragon's mode of propulsion as he is both a flying sky-borne creature and slithering earth-borne one moving above and below the surface of the earth.

A portmanteau neologism, such as 'flythering' or 'slythering', might do service here (though the latter is close to the name of one of the houses of Hogwarts School of Witchcraft and Wizardry in the Harry Potter books). Heaney is not averse to devising portmanteau neologisms for his translation (as we will see later). The dragon in *Beowulf* seems like a sandworm straight out of Frank Herbert's *Dune*, with the added ability to fly. In biblical terms, he is a swarming creature and so comes under the Levitical interdiction against such creatures (as we will see later). Swarming is also a quality that John Milton gives to Satan and Dante gives to 'the souls of the sullen stuck in slime' in the fifth circle of hell (as we will also see later with both).

After Beowulf had ruled 'the wide kingdom' (in Heaney's translation; *Beowulf*, 1999, p. 70), or 'spacious realm' (in Clark Hall's translation; *Beowulf*. . ., 1950, p. 132) of the Geats for fifty years, 'a certain one in the dark nights began to hold sway' (in Tolkien's translation; Tolkien, 2014, p. 77), or 'have power' (in Clark Hall's translation; *Beowulf*. . ., 1950, p. 132), or 'dominate the dark' (in Heaney's translation; *Beowulf*, 1999, p. 70). The dragon is a creature not only of the dark night, but also of the semi-dark twilight, the liminal zone of demi-dark night and demi-light day. He is 'the ancient twilight-foe' in Clark Hall's translation (*Beowulf*. . ., 1950, p. 135). Later in Clark Hall's translation the dragon is more precisely 'the ancient creature who flew by twilight' (p. 158). The dragon is a friend of the twilight, not a foe of the twilight. He is a foe who flies by twilight and night, and a creature of the twilight. Heaney's translation refers to 'the gloaming' (*Beowulf*, 1999, p. 73). This suggests that the dragon is a creature of dusk rather than dawn, of sunset rather than sunrise. This is especially the case when 'he would dart before daybreak' (p. 73) back to 'his den' (pp. 73 and 95) after 'he wreaks havoc on the Geats' (as in Heaney's marginal gloss; p. 73) at night. 'Den' implies that the dragon is a bear-like creature, the monstrous double or doppelganger of Beowulf (whose name means 'bear').

The dragon is thus specifically a creature of fire emerging from his den at sunset. The dragon in Clark Hall's translation 'flies by night, the smooth malicious dragon, burning and wrapped in flame' (*Beowulf*. . ., 1950, p. 135). Tolkien translates these lines as 'the old despoiler wandering in the gloom [. . .] who filled with fire seeks out mounds (of burial), the naked dragon of fell heart that flies wrapped about in flame' (Tolkien, 2014, p. 79). The dragon is a grave robber who plunders burial mounds at night, robs them of their funerary treasures and adds them to the collection he already possesses in his hoard. The dragon is portrayed as a figure of avaricious acquisitiveness, a psycho-pathologically anal sadist who loves acquiring objects, adding them to his collection, possessively owning them and jealously guarding them.

Heaney translates these lines as 'an old harrower of the dark [. . .] the burning one who hunts out barrows,/the slick-skinned dragon, threatening the night sky/ with streamers of fire. . .He is driven to hunt out/hoards underground' (*Beowulf*, 1999, p. 72). In the introduction to his translation, Heaney says of the dragon that 'he is at once a stratum of the earth [. . .] abiding in his underneath [. . .] and a streamer in the air [. . .] fireworking its path across the night sky' (*Beowulf*, 1999, p. xix; Heaney, 2007, p. xxi). The dragon is also a compulsive hoarder and arsonist driven by the compulsion to repeat acts of acquisition and conflagration. Later in Heaney's translation when the dragon is 'hunting for enemies' among 'the humans it loathed' these drives 'drove it to attack' Beowulf (*Beowulf*, 1999, p. 84). The dragon is a hunter, both a human- and a hoard-hunter. The drives of loathing, hunting, hoarding and firing compel and drive the dragon to perform them repeatedly. The dragon in Heaney's translation is a personification, or embodiment, or animalization, of conflagration, a 'dragonflagration', as James Joyce might have put it if the author of *Finnegans Wake* were translating *Beowulf* rather than his Irish compatriot in Seamus Heaney who is equally fond of neologisms (for example, 'scaresomely') and archaisms (for example, 'swinged'). Perhaps Heaney here had Shakespeare in mind, when in *King John* (II, i, 288), he refers to 'Saint George that swing'd the dragon', thus linking the two dragons, and dragon-slayers, albeit anachronistically and reversing their roles.

In Heaney's translation 'the fire-dragon' (*Beowulf*, 1999, p. 74) 'swinged the land, swathed it in flame' (p. 74) and 'scorched the ground' when, 'hot and savage, he kept circling and circling' (p. 73) in the air. The Geats on the ground 'bore the brunt of his brutal assaults/and virulent hate' (p. 73). The fire-dragon practices a scorched-earth policy. He embodies or personifies this practice. Similarly, when 'fiery dragons were seen flying in the air' over Northumbria in north-east England in 793CE they were read as dire portents of Viking raids (as documented by the Anglo-Saxon chronicler; cited by Parker, 2015, p. 9). Just as the theorists of aerial firebombing of cities in World War II turned to the biblical precedent of the fiery destruction of the cities of Sodom and Gomorrah and invoked it explicitly to legitimatize it,[4] so the theorists of the scorched-earth policy and practice in warfare may have had *Beowulf's* dragon in mind, however unconsciously. Or perhaps the dragon's scorched-earth practice was a prophetic precursor of the scorched-earth policy and practice in warfare.

The dragon is not only a creature of fire, night and twilight, but also a brightly shining creature in, or of, the night sky. In Heaney's translation, the dragon with his 'enameled scales' (*Beowulf*, 1999, p. 81) would 'glitter and glide/and show himself off in midnight air' (p. 89) as he 'shimmered forth/on the night air' (p. 95). Draco, or 'the dragon', is a constellation in the northern night sky in which stars are shaped like a dragon. The dragon is an embodiment of Draco, or Draco is the projection of the dragon. The dragon is the star (in a variety of senses) of this part of *Beowulf*, just as Satan was the hero of *Paradise Lost* for William Blake. In this respect, the dragon *is* more *draco* than *dragonitas.* Perhaps Draco and the setting sun are conflated, or conflagrated, to form a monstrous thermonuclear fireball of draconian shape who rains down fiery terror from on high.

Moreover, perhaps the crepuscular conflagration the dragon creates looks like a bushfire sun or forest fire at sunset in which the air above is a sea of flames emitted from the burning bushes and trees below. When in Heaney's translation the dragon 'burn[s] bright homesteads, these was a hot glow/that scared everyone [. . .] Everywhere the havoc he wrought was in evidence' (*Beowulf*, 1999, p. 73). The dragon is a havoc-wreaker. The dragon's destructive conflagration is also a precursor of firebombing in aerial warfare in which the air above the earth was on fire. Where Clark Hall has 'the hateful creature flying in the air' (*Beowulf*. . . , 1950, p. 137) and Tolkien has 'the fell winger of the air' (Tolkien, 2014, p. 81), Heaney has 'the vile sky-winger' (*Beowulf*, 1999, p. 73) sounding like the nickname painted on the fuselage of a fighter-bomber. Heaney also has Beowulf describe the dragon as 'the sky-borne foe' (p. 80). The dragon is a sort of combination of a fighter in a one-pilot air force of one single-engine flying fortress, an airborne flame-thrower and a wing commander of aerial warfare.

The dragon engulfs the elements of earth and air with the element of fire. The distinction between all three, and the proper place for all three is obliterated with fire on the earth and in the air. In the natural order of things according to the philosophy of the four elements, fire should remain below the earth in subterranean lava, or be contained in the hearth of the human home on the earth, or be located in the heavens in the sun and stars. It should not be used as a weapon against humans on the earth. To do so is to violate the elemental order of things. The dragon is a liminal creature on the threshold between the elements and in the twilight between night and day. The dragon is also a liminal creature living on the threshold between the air above the earth and in the earth below the ground. He is a mound monster and a subterranean and super-terrestrial flying serpent. In Clark Hall's translation the dragon is 'the terrible cave-dragon', the keeper of a treasure hoard housed in 'a high stone-barrow' who lives in a cave in 'earth's bosom' (*Beowulf*. . . , 1950, pp. 132, 161 and 171). In Heaney's translation 'the dragon from under-earth' is 'the cave-guard' and 'the hoard-guard' who 'hide[s] in his den' (*Beowulf*, 1999, pp. 73, 80, 82 and 89). The dragon in Tolkien's translation is 'the monstrous guardian' who 'beneath the earth kept those golden treasures' (Tolkien, 2014, p. 83).

The dragon is a creature created by evil out of fire in the earth who lives underground, flies in the air above the earth, breathes out fire, reduces solid structures to dust and hoards treasure in a mound in the ground, whereas humans are creatures created by God out of the dust of the earth (*Genesis* 2: 7) who in Western culture live in houses built on the surface of the earth, walk on the earth, breathe air in and out, control fire in fireplaces, hearths, ovens, stoves, furnaces and engines, and hoard treasure in strongholds built on or under the ground. The dragon mixes the elements and traverses the heights and depths in a violation of the elemental order of things in a way that humans are incapable of. The dragon's monstrosity consists precisely in these violations. He is the sub- and super-terrestrial counterpart living in various liminal zones to the aqua-terrestrial marsh monster and swamp serpent living in the liminal zone in the slime between land and water. The dragon throws into the relief the problematic nature (in two senses) of the philosophy of the elements.

The dragon is a creature of horror and terror. He generates these passions within himself, and he and his treasure hoard elicit them in others. When, in Clark Hall's translation, the slave with his 'sin-perplexed soul' sees the treasure hoard, a 'grisly horror rose up in the stranger' and 'terror held him' (*Beowulf*. . . , 1950, p. 133). Tolkien translates these lines as 'upon the trespasser dire terror fell' (Tolkien, 2014, p. 78). The dragon in Clark Hall's translation is 'the grisly monster' (*Beowulf*. . . , 1950, p. 149), 'a terror to the sons of men' (p. 137), 'the terrible stranger' (p. 148) and 'the vengeful stranger' (p. 155). The slave's sinful soul is figured as monstrous and represents the sinfulness of the dragon, while the dragon is a projection of sinfulness. The slave's soul and the monstrous dragon are mirror images of each other. Horror and terror are reflected from one to the other. Horror is a response to the monstrous (in this case, sinfulness and the dragon), whereas terror is a response to the monumental (in this case, the slave's sight of the treasure). Later in the story terror consists of the dragon's destructive revenge exacted on human settlements for being robbed, and still later in the dragon's and Beowulf's response to each other before they start their fight ('in each one of the hostile pair was terror at the other', in Clark Hall's translation; *Beowulf*. . . , 1950, p. 149).

When the dragon discovers that the slave has robbed him of a goblet he is described in Clark Hall's translation as 'bursting with rage'. He 'meant to requite with fire the theft of the costly drinking bowl' so he 'went forth with flame, furnished with fire' (*Beowulf*. . ., 1950, p. 136). Heaney translates these lines as 'his pent-up fury/at the loss of the vessel made him long to hit back/and lash out in flames' and in a 'fiery blaze' (*Beowulf*, 1999, p. 73). In Clark Hall's translation, the dragon fights Beowulf with 'murderous fire' and 'hostile flames' (*Beowulf*. . ., 1950, p. 150). In Heaney's translation, he is 'the fire-breathing dragon' (*Beowulf*, 1999, p. 84). Fire is the dragon's initial weapon of choice that he wields from a distance coming out of his mouth as he flies above the earth. The result is that 'the whole countryside is consumed by the fire from the dragon's mouth; even Beowulf's stronghold is attacked by the flames', in Clark Hall's gloss on the following passage:

> Then the fiend began to vomit forth flames, to burn the noble dwellings; the gleam of fire blazed forth [. . .] The serpent's warfare was widely visible, the vengeance of the devastator far and near [. . .] he surrounded the people of the land with fire, with flame and burning [. . .] The flaming dragon had wasted with fire the stronghold of the people.
>
> (*Beowulf*. . ., 1950, pp. 137–138)

Tolkien translates the last sentence as 'the flaming dragon [. . .] with glowing fires had crushed to ruin the stronghold of the folk, the guarded realm' (Tolkien, 2014, p. 81). Heaney translates this sentence as 'the fire-dragon [. . .] reduced/forts and earthworks to dust and ashes' (*Beowulf*, 1999, p. 74). The external wasteland the dragon produces is a manifestation of the interior 'wildness [. . .] in the dragon' (p. 84). Inner and outer express and reflect each other. The feral monster produces a feral quaking zone in contrast to the native quaking zone of the marsh and swamp (see Giblett, 2009, chapters 1 and 2).

The dragon's mound is described in Heaney's translation as surrounded by 'that desert waste' (*Beowulf*, 1999, p. 73). Some commentators on the biblical creation story of *Genesis* 1: 2, such as Westermann (1984, pp. 76 and 103), prefer to translate the Hebrew *tohu wabohu* as 'desert waste', rather than as 'primeval chaos' or 'formless and empty' as in the older and standard translations. The dragon as a creature living in a desert waste thus can be read as a survivor of, or throwback to, the first day of creation just after God created the heavens and the earth, but before He divided the heavens, waters and earth from each other, and light from darkness. Similarly in the creation story in *The Legend of Sigurd and Gudrún* (at least in Tolkien's version) from 'an abyss yawning/[. . .] dread shapes arose/from the dim spaces/[. . .] friends of darkness,/[. . .] unbegotten,/out of ancient void/[. . .] on the margin of the world', including 'the slumbering serpent/in the sea' and 'the deep dragon' (Tolkien, 2010, pp. 59–63). Unlike the monsters Grendel and his mother, who were created after the Fall, and so after the creation of the animals (as related earlier in *Beowulf* and as discussed in the previous chapter), the dragon is a creature of the desert waste of primeval chaos created before the animals. The dragon is a creature of primeval chaos; in a sense, he *is* primeval chaos. All three monsters were not created by God and are not part of God's creation. The dragon's desert wastelands ravaged by fire and reduced to ruins, dust and ashes are also an uncanny prophecy or disturbing echo of the ruined wasteland of German cities in the aftermath of World War II firebombing (see Giblett, 2009, chapter 5).

The dragon is an orally sadistic monster who consumes with fire emitted from his mouth and who rains down terror from the sky. Unlike the monstrous crocodile who kills in water by drowning its victims in a death roll, marinating and tenderizing them in water, and then consumes them by ingesting and digesting them, the monstrous dragon consumes by heaving up, spewing out, spouting, vomiting or belching fire and flames that devastate and destroy. The slimy crocodile is a water monster; the earth dragon a fire monster. Tolkien's translation has 'the invader did now begin to spew forth glowing fires' (Tolkien, 2014, p. 80). Clark Hall glosses 'the dragon [. . .] spewing flame' (*Beowulf*. . . , 1950, p. 148), and later he also glosses how 'the reptile belches forth more flame' (p. 149). Whether the dragon is a reptile is another question. As he flies and as he is living in the age of humans when there are no flying reptiles, he may be a survivor of, or throwback to, the age of dinosaurs when there were flying reptiles, and no humans.

Unlike dinosaurs, though, the dragon breathes out fire through his mouth. In Heaney's translation 'the dragon [. . .] belch[ed] out flames' (*Beowulf*, 1999, p. 73), 'spouted deadly flames' (p. 81) and was 'heaving up fire' (p. 84). Spewing, spouting, vomiting, belching and heaving up are all oral emissions of solid, liquid or gaseous matter (or various combinations thereof), whereas fire is not usually emitted orally, except by dragons. The dragon is the exception who exhales fire orally (and not nasally). The ability to emit fire orally is one of the monstrous features of the dragon that defies and deviates from the divinely ordained natural order of things.

The dragon's fiery breath is not only hot and incendiary, but also venomous and deadly. In Clark Hall's translation 'the monster's breath' is described as 'the hot vapour of battle' (*Beowulf*. . . , 1950, p. 148) that produces 'deadly fumes' (p. 153). When the dragon wounds Beowulf with his sharp fangs or teeth, 'the deadly venom seethed within his breast, — poison within him' (p. 156). Heaney's translates this as 'the deadly poison suppurating inside him' (*Beowulf*, 1999, p. 85). Before Beowulf fights the dragon he says he will be 'meeting molten venom/in the fire he breathes' out (p. 80). After he is dead the dragon is described in Heaney's translation as 'that poison-breather' (p. 89). The dragon is a serpent who breathes out venomous fumes from his mouth and injects venom with his fangs. Both the dragon's exhalation and injection of venom are deadly. Heaney's translation emphasizes that the dragon is a fire-breathing, orally sadistic, poisonous serpent of the volcanic underground. He is an animalization not only of fire, but also of volcanic poison in the liquid form of venom and in the gaseous form of poisonous vapour and fumes.

The dragon breathes out fire through his mouth and presumably breathes in air through his nose. Somewhere and somehow within the dragon's unique physiology, air is transformed into fire in a mysterious biochemical or alchemical process, or both. The dragon's anatomy is an inversion of usual animal anatomy as his mouth exhales fire, just as the anus excretes excrement and emits flatulence. Presumably the dragon eats with his mouth and ingests food through it, as well as excretes and farts through his anus, but he also expels fire from his mouth. The dragon is an orally and anally sadistic monster. He is anally sadistic as he is a hoarder who loves possessing treasures and he is orally sadistic as he breathes out destructive fire and venom. In Tolkien's version of *The Legend of Sigurd and Gudrún* Fafnir the dragon has both 'fire and venom'. He came forth 'fire his breathing' with 'mists of poison' (Tolkien, 2010, pp. 104 and 108).

Rather than the dragon himself consuming orally and sadistically by ingesting, the dragon's fire consumes orally and sadistically, such as in 'the horror' of Beowulf's own home being 'swallowed by the waves of fire' (as in Clark Hall's translation; *Beowulf*. . . , 1950, pp. 137–138), or 'burnt to a cinder' (as in Heaney's translation; *Beowulf*, 1999, p. 74). The destruction of 'his own home [. . .] threw the hero/into deep anguish and darkened his mood' as Heaney's translation goes on to relate (p. 74). 'The war-king planned and plotted his revenge' against 'the sky-plague', 'the scourge of the people', as Heaney's translation describes the main protagonists of this part of *Beowulf* (pp. 72–74). The war-king hero has been infected with the darkness, deep hatred and virulent disease of the sky-plague scourge. The dragon is an air-borne pathogen, a winged draconian grim reaper, death writ large and on the wing, a flying *momento mori*, and an embodiment and instrument of *Genesis* 3: 19 that 'you were made from dust, and to dust you will return'. In his introduction to his translation Heaney says that 'dragon equals shadow-line, the psalmist's valley of the shadow of death' (*Beowulf*, 1999, p. xix; Heaney, 2007, p.xxi). The dragon adumbrates the liminal zone of death.

Moreover, the dragon embodies and inflicts death; he, in a sense, *is* death. In Heaney's translation, he 'would leave nothing alive in his wake' (*Beowulf*, 1999, p. 73). The dragon is a death-dealing airship and airborne 'nightmarish destroyer' (p. 89). He is a flying battleship in a technological anachronism for the eighth century of the Common Era. The two figures of 'the hero' and 'the war-king', on the one side of the battle, and 'the dragon' and 'the sky-plague', on the other, confront and reflect each other across the metaphysical deep, dark divide of mutual hatred in a Manichean world of good versus evil, life versus death. They mirror each other across this divide before they grapple each other physically in battle and end up destroying each other (as Heaney's translation emphasizes; *Beowulf*, 1999, p. 89). The physical battle between them is an outworking and manifestation of the metaphysical, Manichean divide between them.

The monstrous dragon exhales orally sadistic consuming fire that engulfs his targets and victims in waves and leaves death in his wake. The dragon's 'surging fire' in Clark Hall's translation (*Beowulf*. . . , 1950, p. 154) is like a powerful tidal force in a sea of flames. Many victims of the allied firebombing of German cities in World War II described the devastating aftermath as a 'sea of flames', or *Flammenmeer* in German (see Giblett, 2009, chapter 5). The dragon produces a sea of flames; he, in a sense, *is* fire, the embodiment and destructive application of deadly fire. He is 'the flaming dragon' (in both the Clark Hall translation; *Beowulf*. . . , 1950, p. 138 and Tolkien, 2014, p. 81). In Heaney's translation, the dragon is 'the monster' of the 'flaming depths [. . .] swaddled in flames' (*Beowulf*, 1999, pp. 80–81). The dragon is born of fire under the earth, is wrapped in fire and exhales fire on the earth. The dragon swaddled or 'wrapped in flame' is the anti-Christ who parodies *Luke* 2: 12 in which the baby Jesus is 'wrapped in swaddling clothes'.

In his lecture on *Beowulf*, Tolkien (2006, p. 23) concludes his comments on the topic of fire that 'the dragon wields a physical fire' and not 'the fiery darts of the wicked'. The dragon is a creature of physical fire and of moral wickedness, and not of spiritual fire and satanic evil. Tolkien calls the dragon 'the serpent' in between these two comments, alluding to the satanic and wicked serpent in the biblical Garden of Eden. Clark Hall's translation refers to 'the evil beast' (*Beowulf*. . . , 1950, p. 136), 'the dread malicious spirit' (p. 154), 'the dread fiery dragon' (p. 155) and 'the fiery dragon, terribly bright' (p. 171). Tolkien's translation refers to 'the serpent, alien creature fierce and evil' (Tolkien, 2014, p. 91) and notes 'him do earth dwellers greatly dread' (p. 79). In Heaney's translation 'people on the farms are in dread of him' (*Beowulf*, 1999, p. 72). The dragon is an alien, dreadful, wicked, serpentine, supernatural and a pre-Christian or non-Christian evil spirit. The treasure the dragon hoards and guards is 'heathen gold' in both Clark Hall's (*Beowulf*. . . , 1950, p. 135) and Heaney's translations (*Beowulf*, 1999, p. 72) and 'pagan treasure' in Tolkien's (2014, p. 78).

The dragon is an orally sadistic monster who not only breathes out fire, but also bites with his sharp fangs or teeth and injects deadly venom into his victims through puncture wounds. The finale of the fight to the death between the dragon and Beowulf is gruesome as 'the dread fiery dragon [. . .] enclosed his [Beowulf's] whole neck between his sharp teeth; he was bathed in life-blood – the gore gushed out in streams' in Clark Hall's translation (*Beowulf*. . . , 1950, p. 155). Tolkien's translation refers to 'his sharp bony teeth' (Tolkien, 2014, p. 91). In Heaney's translation, the dragon 'clamped sharp fangs into his [Beowulf's] neck' (*Beowulf*, 1999, p. 85). The dragon is an orally sadistic monster with sharp teeth and fangs suited to efficiently inflicting painful wounds and injecting venom.

Beowulf eventually dies as much from the poison injected by the dragon's deadly fangs or teeth as from loss of blood from the wounds inflicted by them. In Tolkien's translation Beowulf 'died a monstrous death' in 'that monstrous sight to see' (Tolkien, 2014, p. 101), or in 'the uncanny scene' in Heaney's translation (*Beowulf*, 1999, p. 95). Given the strong association between the monstrous and the uncanny, and their association with death and dying, 'uncanny scene' and 'monstrous sight' seem more appropriate than Clark Hall's literal or cognate translation of the Old English 'wundur scéawian' (Heaney, 2007, p. 204) as 'the wondrous sight' (*Beowulf*. . . , 1950, p. 171).

The dead dragon in Tolkien's translation is 'more strange' (Tolkien, 2014, p. 101) and in Heaney's translation 'far stranger' than the monstrous sight or uncanny scene of Beowulf's death: 'the serpent on the ground, gruesome and vile/ lying facing down. The fire-dragon/was scaresomely burnt, scorched all colours' (*Beowulf*, 1999, p. 95). The fire-dragon has become a rainbow serpent, an animalization of the rainbow. Draco, the constellation dragon shining brightly and flying in the night sky, is reduced to the dead rainbow serpent brought to earth and prostrated on the ground. Tolkien's translation of these lines emphasizes that, 'grim to see, deadly-hued, the flaming dragon had been scorched with his own glowing fires' (Tolkien, 2014, p. 101).

As the dragon also coils, twists and writhes, he probably burnt himself after Beowulf wounded him. In Heaney's translation, when the dragon discovers he has been robbed 'he rippled down the rock, writhing with anger' (*Beowulf*, 1999, p. 73). The dragon is a water-like, serpentine creature. Before fighting Beowulf, Clark Hall's translation describes how 'the coiled serpent' (*Beowulf* . . . , 1950, p. 162) with 'the coiled creature's heart' (p. 148) 'quickly coiled itself together (149), then came towards Beowulf 'fiery and twisted' (p. 149). In Tolkien's version of *The Legend of Sigurd and Gudrún* Fafnir the Dragon is described as 'coiling' (Tolkien, 2010, p. 108). In Heaney's translation of *Beowulf* 'the outlandish thing/writhed and convulsed and viciously' attacks Beowulf. Beowulf holds his ground so 'the serpent looped and unleashed itself' (*Beowulf*, 1999, p. 81) against Beowulf. Perhaps the dragon stretches out to his full length of fifty feet so the monster is of truly monstrous proportions. After Beowulf 'cuts the reptile asunder' in Clark Hall's translation (*Beowulf* . . . , 1950, p. 155), perhaps the dragon/serpent/reptile writhed in agony so much that he/it scorches and burns himself/itself. Perhaps the dragon died as much from self-inflicted burns as from Beowulf's dagger. Both Beowulf and the dragon die dual-deaths in their deadly duel with two possible causes for both of their deaths. The dragon dies by dismemberment and/or self-immolation and Beowulf from blood loss and/or poisoning.

St George's swamp dragon

The death of the dragon in *Beowulf* in the eighth century CE by no means meant the death of the dragon in Western culture. Quite the contrary was the case. In the eleventh and twelfth centuries CE the legend of St George and the dragon started to circulate widely in Europe. Its success in England was partly attributable to *Beowulf* and its dragon. In one version of the St George legend 'a terrifying dragon' is an inhabitant of a lake and an instrument of God's wrath. God sent the dragon 'which devoured the inhabitants' of the nearby city ruled by 'a godless emperor' (Collins, 2012, p. 90). St George kills the 'huge and awful' dragon with his sword.

St George is shown killing the dragon with his sword in probably the most famous depiction of them all – Benedetto Pistrucci's design of Saint George and the dragon on many mintings of British gold sovereigns and half-sovereigns dating from 1817. Collins (2012, p. 203) describes this design in his exhaustive catalogue of all things related to St George and the dragon as 'one of the most beautiful designs ever to appear on a coin'. It may be a beautiful design, but it depicts the gruesome and horrible slaying of the dragon in an act of gross animal cruelty. The beauty of the design hardly mitigates or overcomes the ugliness of the dragon-slaying.

For Ingersoll (2014, p. 194), writing in the 1920s, the two main protagonists of the legend are 'the proper, most eminent Saint George, and his most celebrated and distinguished of all dragons'. Unlike the dragon of *Beowulf* who is an earth monster, the dragon of the legend of St George is a water monster, or more precisely a marsh monster, like Grendel and his mother in *Beowulf*, and/or a swamp serpent. The dragon in general for Ingersoll (2014, p. 15) has 'only one definable characteristic – association with and control of water'. This is certainly not the case with the dragon of *Beowulf*, who has a variety of defining characteristics, such as living in the earth and association with, and control of, fire. In Ingersoll's summary of the St George legend 'a loathsome monster' inhabits 'a pond' of 'dark water'. Like the dragon of *Beowulf*, 'the exhalations of its breath poisoned all who came near'. St George meets 'the horrible fiend' and 'transfixed the beast' with his lance (Ingersoll, 2014, pp. 194–195). Most depictions of Saint George and the dragon show him killing the dragon with his lance.

In the thirteenth century CE *The Golden Legend* for Collins (2012, p. 91) 'popularized the legend of George and the dragon in the West [. . .] The story may have been particularly well-received in England because of the legend of Beowulf in Anglo-Saxon literature' dating from the eighth century containing the English Ur-story of a dragon. Caxton published an English translation of *The Golden Legend* in 1483. It describes 'a stagne or a ponde lyke a see wherein was a dragon whyche envenymed all the contre' (cited by Collins, 2012, p. 93). 'Envenomed' is largely lost from English today (Robert Coover (1991, p.705) uses it in 'The Dead Queen'), but is certainly precise when considering the activities of the dragon of *Beowulf*. The dragon of St George follows suit. The dragon in *The Golden Legend* is not merely a water monster, but specifically a swamp serpent or marsh monster. This is certainly the case for Baden-Powell, the founder of the Boy Scouts, for whom the dragon 'lived in a swamp' (cited by Collins, 2012, p. 128). Similarly, Sabine Baring-Gould's *Lives of the Saints* of 1872 recounts the legend in which 'a pond [is] infested by a monster' and how St George sees 'the monster rising from the marsh' (cited by Collins, 2012, p. 159).

The dragon in Christianity, Collins (2012, p. 92) concludes, 'symbolized the Devil, heresy, paganism and ultimately anything destructive of [Christian?] society [. . .] Many a dragon has the curled and twisted tail of a serpent, suggestive of evil as old as Eden'. The dragon in Boehm's statue of Saint George and the dragon outside the State Library of Victoria in Melbourne has a forked tongue, which is certainly suggestive of a serpent and of deception. Whether this is also suggestive of evil as old as Eden is another question. Geoffrey Serle (1971, p. 287) might concur with Collins when he argued that this statue signifies 'the triumph of good over evil'. Yet the dragon as a creature of primeval chaos as depicted in *Genesis* 1: 2 may be older than Eden and not suggestive of evil at all. A twisted tail and forked tongue are suggestive of a serpent, but are not necessarily suggestive of evil, as old as Eden or not, and only with the hindsight of the biblical story of the Fall.

Collins (2012, p. 93) goes even further to suggest that the dragon's 'poisonous breath may be taken to be heretical preaching'. Unlike the dragon's poisonous breath or venomous vapour that could kill (as in *Beowulf*), heretical preaching cannot, though it might get the heretical preacher burnt at the stake so in a sense it could kill – by fire, like the dragon's fiery breath. The dragon was a salutary warning that heretical preaching was dangerous and could kill.[5]

Malory's marvellous dragons

In the fifteenth century of the Common Era Caxton published another story with a veritable bestiary of monsters. Thomas Malory's *Le Morte d'Arthur* relates the much older legends of King Arthur and his knights of the Round Table dating from the fifth century CE. In chapter 19 of Book 1 Arthur dreamed 'a marvellous dream' in which 'there was come into this land griffins and serpents [. . .] and the strangest beast he ever saw [. . .] the Questing Beast' (Malory, *I*, 1969, p. 46). Chapter 12 of Book 9 describes how 'the Questing Beast [. . .] had in shape a head like a serpent's head, and a body like a leopard, buttocks like a lion, and footed like an hart' (Malory, *I*, 1969, p. 401). The hybrid anatomy of the questing beast is like the dragon in some stories.

King Arthur later dreamed about a dragon. Chapter 5 of Book 5 describes how Arthur:

> dreamed a marvellous dream: him seemed that a dreadful dragon did drown much of his people, and he came flying out of the west, and his head was enamelled with azure, and his shoulders shone as gold, his belly like mails of a marvellous hue, his tail full of tatters, his feet full of fine sable, and his claws like fine gold; and an hideous flame of fire flew out of his mouth, like as the land and water had flamed all of fire.
>
> (Malory, *I*, 1969, p. 172)

This dragon is a metallic, fire-breathing monster very similar in his anatomy and physiology to Beowulf's dragon of *Beowulf.*

An even more hideous monster comes out of the east in Arthur's dream in hot pursuit of the dragon out of the west:

> After, him seemed there came out of the orient, a grimly boar all black in a cloud, and his paws as big as a post; he was rugged looking roughly, he was the foulest beast that ever man saw, he roared and romed so hideously that it were marvel to hear. Then the dreadful dragon advanced him and came in the wind like a falcon giving great strokes on the boar, and the boar hit him again with his grizzly tusks that his breast was all bloody, and that the hot blood made all the sea red of his blood. Then the dragon flew away all on an height, and came down with such a swough, and smote the boar on the ridge, which was ten foot large from the head to the tail, and smote the boar all to powder both flesh and bones, that it flittered all abroad on the sea.
>
> (Malory, *I*, 1969, pp. 172–173)

The monstrous boar or bear (in Keith Baines 'rendition' (Malory, 2001, p. 87)) of the east is more hideous than the monstrous dragon of the west.

Also unlike Beowulf and the dragon in *Beowulf* who are Manichean mirror images of each other, Arthur's dragon is himself, or an allegorical version of himself, at least according to the dream interpreter:

> And therewith the king awoke anon, and was sore abashed of this dream, and sent anon for a wise philosopher, commanding to tell him the signification of his dream. 'Sir,' said the philosopher, 'the dragon that thou dreamedst of betokeneth thine own person that sailest here, and the colours of his wings be thy realms that thou hast won, and his tail which is all to-tattered signifieth the noble knights of the Round Table; and the boar that the dragon slew coming from the clouds betokeneth some tyrant that tormenteth the people, or else thou art like to fight with some giant thyself, being horrible and abominable, whose peer ye saw never in your days, wherefore of this dreadful dream doubt thee nothing, but as a conqueror come forth thyself.'
>
> (Malory, *I*, 1969, p. 173)

This dream could be read as a premonition or prophecy of the later contests of Britain versus the Roman Empire, King Arthur versus the Giant or Emperor Lucius, London versus Rome, or as an allegory of west versus east, Christianity versus Islam, capitalism versus communism – all depending on the era.

Not to be outdone, the greatest knight of them all, Sir Launcelot du Lake, gets in on the act of dragon- or serpent-slaying in chapter 2 of Book 11 of Malory's *Le Morte d'Arthur*. The serpentine dragon is another fire-breathing monster who lives in a tomb, and so is a creature of both death and the underworld. When Sir Launcelot lifted up the lid, 'there came out an horrible and fiendly dragon, spitting fire out of his mouth'. The dragon is also a fiend, a sinful creature of spiritual evil. Sir Launcelot 'drew his sword and fought with the dragon long, and at the last with great pain [. . .] slew that dragon' (Malory, *II*, 1969, p. 190). Sir Launcelot thus joins the noble company of dragon-slayers that Beowulf was later to join. He also joins the Christian company of evil-destroyers.

Colonna's serpentine dragon

In 1499 Franceso Colonna in his *Hypnerotomachia Poliphili* has Polyphilo dream of 'a frightful and terrible dragon' with a variety of typical features:

> Its triple tongue trembled in its jaws, which were [as] full as a comb with sharp, serrated teeth. Its fat, scaly body slithered over the tiled pavement, and as it glided, its wings slapped its furrowed back, while its long tail wound itself in serpentine fashion into tight, unstable knots.
>
> (Colonna, 1999, p. 62)

The tongue of this dragon is not forked into two segments (as with Boehm's statue), but tripled.

The mouth of the dragon is an object of horror and fascination as 'the monster was spewing out foul smoke' (p. 62) and 'the sharp teeth of the venomous dragon' of 'its crushing, saw-toothed jaws' were indicative of its 'voracious appetite' (p. 63). There is no doubt that this dragon is an orally sadistic monster. Polyphilo fears 'the fierce appetite and violent maw of this terrible dragon' (p. 64) and of being eaten alive: 'Alas, what if I am swallowed whole to rot inside its foul, filthy and faecal entrails, to be afterwards ejected by an unthinkable exit?' (p. 64). The dragon enacts fears of an oral and excremental nature, of being consumed alive in the mouth of an orally sadistic monster, of being transformed into rotting matter in the belly of the beast and of being excreted as waste matter from his fundament. The journey into the belly of the beast is an inner journey into 'deep darkness' in which Polyphilo imagines that 'my human body and flesh shall nourish this terrible beast' (p. 64). Rather than the body and flesh of beasts nourishing humans, 'an atrocious and monstrous reversal', not only of fortune might occur as Polyphilo says, but of the order of things. The eater is eaten.

Milton's draconian Satan

John Milton writing in the seventeenth century inherited the rich tradition of monsters and injected it into his retelling of the biblical story of how paradise was lost. Satan in Milton's *Paradise Lost* is a swamp serpent and marsh monster, as is the dragon in some versions of Saint George and the dragon. In fact, Milton's Satan is a dragon and more precisely a miasmatic marsh monster. He is described as 'involved in rising mist' (*Paradise Lost*, IX, 75) and as 'wrapped in mist/Of midnight vapour' who glides 'obscure' (IX, 158–9) 'like a black mist low creeping' (IX, 180). Satan is a miasmatic marsh monster because, as a fallen angel, he has been 'mixed with bestial slime,/This essence to incarnate and imbrute' (IX, 165–6). Milton's Satan is not merely a creature of water, but specifically a creature of slime. Satan 'personifies', or more precisely 'monstrifies', slime and wears as his proper wardrobe the miasma which arise from it. He monstrifies slime because he was 'dragon grown. . .engendered in the Pythian vale on slime' (X, 530). He is born from slime and made of slime; he is slime by birth and slime by nature.

With the dragon Satan coming out of slime and being made of slime, Milton may have been following in Dante's slimy footsteps. In the seventh Canto and the fifth Circle of Dante's *Inferno* (and not the tenth level, as the Mayor of New York surmises incorrectly in *Ghostbusters II*) the condemned souls of the sullen are stuck in the slime of the Styx, transformed from river to marsh by Dante, complete with swampy black water (Dante, 1996, pp. 57–58). The slime of Hell is the product of the marsh or swamp of Hell. Milton's Satan is a creature of the slime of the swamp of Hell.

Slime has been associated with the excremental, especially when the turd is not hard and dry, nor soft and wet, but in between both sets of categories (as I suggest elsewhere; see Giblett, 1996, chapter 2). For Milton it also seems, as Stephen Greenblatt (1982, p. 12) argues for Martin Luther, that 'the Devil dwells in excrement'. As a slimy swamp serpentine dragon, his mode of propulsion is not restricted to any one kind, but varies according to the (wet)landscape over which he is travelling. As he does not stick to one mode of propulsion he is aberrant: 'so eagerly the Fiend/ O'er bog or steep, through strait, rough, dense, or rare,/With head, hands, wings, or feet pursues his way,/And swims or sinks, or wades, or creeps, or flies' (II, 947–50). Satan as a dragon is even a flying miasmatic marsh monster and swamp serpent. He is monstrous because he combines the physical characteristics of birds (he flies), fish (he swims), reptiles (he creeps), and quadrupeds or bipeds (he sinks or wades). In a word, he swarms.

Satan comes 'swarming now/With complicated monsters, head and taile,/ Scorpion and Asp' (X: 522–524). The dragon Satan is a swarming monster and so comes under the biblical interdiction of *Leviticus* 11: 41 that 'every swarming thing [. . .] is an abomination'. Swarming creatures, Mary Douglas has commented, are:

> both those which teem in the waters and those which swarm on the ground. Whether we call it teeming, trailing, creeping or swarming, it is an interminable form of movement [like slime] [. . .] 'swarming' which is not a mode of propulsion proper to any particular element, cuts across the basic classification [like the wetland]. Swarming things are neither fish, flesh nor fowl. Eels and worms inhabit water, though not as fish; reptiles go on dry land, though not as quadrupeds; some insects fly, though not as birds. There is no order in them [. . .] As fish belong in the sea, so worms belong in the realm of the grave with death and chaos.
>
> (Douglas, 1966, p. 56)

Milton's Satan is a worm that belongs in the tomb/womb of the swamp with slime; *Beowulf's* dragon is a worm that belongs in the tomb/womb of the grave with death. Both are creatures of the fertile depths of the earth and are projections of fears of an excremental nature.[6]

Anatomy and physiology of the dragon

The dragon assumes a variety of guises and shapes, and is even a shape-shifter who serves a range of functions depending on the role he is asked to perform in the story in which he appears. Attempts to codify the anatomy and physiology of the dragon begin from a small base and neglect one or another feature found in the literature. In his study of the dragon, Ingersoll cites 'Professor Smith' for whom 'the substratum of its [the dragon's] anatomy consists of a serpent or a crocodile, usually with the scales of a fish for covering, and the feet and wings, and sometimes also the head, of an eagle, falcon, hawk, and the fore-limbs and sometimes the head of a lion' (Ingersoll, 2014, p. 27). The dragon, in other words, is an aquatic and amphibious feline raptor at home in the water, on land and in the air – in short, an impossibility! As the dragon invariably breathes out fire, he is at home in this element too – an equal impossibility. Smith neglects to mention fire-breathing (and Milton's dragon is not fire-breathing) as a vital standard feature of the dragon's physiology, perhaps as he was only concerned with the anatomy of the dragon. The dragon is a creature of all four elements and traverses all four of them. Part of his monstrosity consists in this violation of the order of things that makes him into a creature of chaos who does not observe or adhere to the proper order of things with the separation of the elements.

There is a biblical warrant for both Milton's and Smith's anatomy of the dragon and for acknowledging the fire-breathing physiology of the dragon. In John's vision on Patmos recorded in the biblical book of *Revelation* he sees what Ingersoll (2014, p. 131) calls 'dragon-horses' whose heads 'were as the heads of lions; and out of their mouths issued fire and smoke and brimstone [. . .] their power is in their mouth, and in their tails: for their tails were like unto serpents, with heads, and with them they do hurt' (*Revelation* 9: 17 and 19). Like the dragon in *Beowulf*, the dragon of John's apocalyptic vision is an orally sadistic monster who inflicts pain with fire from his mouth. But not for long as 'there was war in heaven: Michael and his angels fought against the dragon [. . .] the great dragon was cast out, that old serpent, called the Devil, and Satan' (*Revelation* 12: 7 and 9; see also 20: 2). The dragon, born out of the war of primeval chaos (or *Chaoskampf*) in *Genesis* 1: 2, is thrown out of heaven in the war with the angels as recorded in *Revelation* 12: 9.

Smith's anatomy of the dragon excludes not only fire, but also some of the features of Giambattista Basile's dragon of 1634. About thirty years before Milton and his dragon Satan, Giambattista Basile (1992, p. 42) published a fairy tale in which 'a seven-headed dragon' appears. This dragon has a similar hybrid anatomy to his predecessors and successors as this dragon was 'the most terrible that had ever been seen in the world. It had the crest of a cock, the head of a cat, eyes of fire, the jaws of a race-hound, the wings of a bat, the claws of a bear and the tail of a serpent'. This dragon is an orally sadistic pagan monster who consumes by ingestion (like the dragon of the St George legend and not by fire, unlike the dragon of *Beowulf*) as it 'swallows a Christian every day'. Like the legend of St George and the dragon, a princess in Basile's account is 'to be swallowed and devoured by this horrible beast'. Like the dragon of *Beowulf*, Basile's dragon comes out from 'the depths of a great cavern'.

The anatomy of later dragons is equally bizarre and deviates similarly from standard animal anatomy. Two and a half centuries after Basile's monstrous dragon, a good and wise dragon in the form of a griffin, but with only two front legs, appeared in a story by Frank Stockton. For Ingersoll (2014, p. 17) 'griffins (perpetuated in the Greek *gryphon*) [are] perhaps the most primitive incarnations of the dragon'. They are also long living as they are preserved in much more modern incarnations. In 1885 Stockton (1992, p. 462) published a fairy tale about the griffin in which a sculptor:

> carved in stone the figure of a large griffin [. . .] the image was not a pleasant one to look at. It had a large head, with enormous open mouth and savage teeth. From its back arose great wings, armed with sharp hooks and prongs. It had stout legs in front, with projecting claws, but there were no legs behind, the body running out into a long and powerful tail, finished off at the end with a barbed point. This tail was coiled up under him [. . .]

In other words, the sculpture of the griffin has the tail of a serpent with the sting of a scorpion in it, another piece of hybrid anatomy. The sculpture of the griffin is a lifelike depiction of the actual flesh and blood griffin of the story. Despite his dangerous barb-pointed tail and his appearance as an orally sadistic monster, the griffin of Stockton's fairy tale is good and wise, unlike most dragons in literature, and better than most humans in the story. He defies Collins' assertion that 'the curled and twisted tail of a serpent [is] suggestive of evil as old as Eden'. Like the dragon of *Beowulf*, the griffin lives in 'the dreadful wilds' in 'the dismal cave he called his home' (Stockton, 1992, p. 471). Yet whereas the *Beowulf* poet is deadly serious, Stockton treats these descriptors of the griffin's abode and its locale with a degree of ironic detachment, if not regarding them as clichés.

Tolkien's earth dragon

In *The Hobbit* Tolkien's earth dragon, Smaug, follows in the footsteps of many previous dragons. He comes from 'the Withered Heath where the great dragons bred' (Tolkien, 2013, p. 33). 'Bred' implies that these dragons engaged in profligate animal reproduction in the past and that they do not breed in the present. Perhaps Smaug is the last of a long line. He certainly ends up being the late Smaug. On the frontispiece map 'the Withered Heath' is captioned as 'whence came the Great Worms' (p. 6). The first definitive characteristic of 'Old Smaug' presented in *The Hobbit* (besides the fact that he is old) is that he is an orally sadistic monster as he has a reputation for 'devouring [. . .] many [. . .] dwarves and men [. . .]' (p. 34). Indeed, Thorin the dwarf says Smaug is 'a most specially greedy, strong and wicked worm' (p. 38). It is taken for granted that 'dragons steal gold and jewels, you know [. . .] and they guard their plunder as long as they live', Thorin explains (pp. 37–38). Like the desert waste made by *Beowulf's* dragon, in *The Hobbit* dragons made 'all the general waste and destruction' (p. 38). Perhaps this is the area indicated on the frontispiece map as 'The Desolation of Smaug'.

Smaug makes 'a noise like a hurricane' and has 'a spout of flame' (p. 38). Smaug later boastfully describes his own anatomy and accoutrements to Bilbo in some detail: 'my armour is like tenfold shields, my teeth are swords, my claws spears, the shock of my tail a thunderbolt, my wings a hurricane, and my breath death!' (p. 274). Smaug does not have fangs, nor does he have venom, unlike *Beowulf's* dragon on both counts. His fiery breath alone (exhaled from both his nose and mouth) means death. Nor does have a sting in his tail like a scorpion, but thunderous electromagnetic power in his stormy tail like lightning. Smaug sounds like one of the Apocalyptic 'dragon-horses'. He also sounds like a parody, or diabolical counterpart, of the Holy Spirit. On the day of Pentecost in the *Acts of the Apostles* (2: 2–3) the Holy Spirit manifests itself when 'suddenly there came a sound from heaven as of a rushing mighty wind' and 'cloven tongues like as of fire sat upon' the apostles. Smaug is an unholy spirit as in 'his wicked inside' (p. 271) 'he had a wicked and a wily heart' (p. 272).

Unlike many previous dragons, Smaug breathes out fire from both his nose and his mouth when 'the nostrils sent forth fire and vapour' (p. 275), 'the dragon spouted terrific flames' (p. 275) and 'fire leapt from the dragon's jaws' (p. 299).

He wears 'a waistcoat of fine diamonds (p. 274). 'His belly glittered white with sparkling fires of gems' (p. 301), except for one place in 'the hollow of his left breast' (p. 300), which Bilbo observes (p. 274) and which later proves the dragon's undoing.

In the final battle Smaug commences his fiery assault on the town of Esgaroth with alacrity: 'before long, so great was his speed rushing towards' the town that he was seen as 'a spark of fire [. . .] growing ever huger and more bright' (p. 298). Smaug not only attacks with fire, but also with noise for 'the roar of Smaug's terrible approach grew loud'. Smaug's aerial attack instils terror from above. In the period in which Tolkien was writing, dive-bombing planes were starting to do so too with screaming engines and shrieking bombs. Tolkien finished writing and published *The Hobbit* during the Spanish Civil War. Beginning with the first aerial bombing by Italians of Turkish troops in 1911 in North Africa, German planes infamously bombed the Spanish town of Guernica and its civilian population in April 1937. *The Hobbit* was published in September of that year. If Tolkien was not aware of bombing in the early Spanish Civil War, perhaps he had seen and heard enough when he was serving as a Lieutenant on the Western Front in World War I and when aerial bombing was in its infancy to know how terrifying and destructive it could be and could imagine how terrifying and destructive it might become. He was certainly prescient about firebombing in World War II.

Smaug does one pass flying over the town, and then:

> roaring, he swept back over the town. A hail of dark arrows leapt up and snapped and rattled on his scales and jewels, and their shafts fell back kindled by his breath burning and hissing into the lake. No fireworks you ever imagined equalled the sights that night. At the twanging of the bows and the shrilling of the trumpets the dragon's wrath blazed to its height, till he was blind and mad with it.
>
> (Tolkien, 2013, p. 298)

Smaug has a number of typical features of dragons. He has scales; he breathes out fire; he flies by night; and he produces an unimaginable display of fireworks. When he is fatally shot in his weak spot with Bard the bowman's 'black arrow', 'Smaug shot spouting into air', like a giant Roman candle, before he 'crashed down from on high in ruin' (p. 301), both his own ruination and the town's as he fell on it and ruined it. Ruins are a notable feature of the aftermath of firebombing.

Dragons perform a number of different functions in the stories in which they play a role. They are expressions and reflections of the cultural concerns of the times in which they are imagined. They also have some common features that persist over time. Generally, dragons are monsters who cross the four elements of earth, air, fire and water. In most stories to date they are evil. They are rarely good. In biblical terms, they are swarming creatures and an abomination. In environmental terms, earth dragons are good and wise keepers or manifestations of the precious metals and minerals of the earth. Humans steal minerals and metals greedily and ungratefully from the earth/dragon, portray dragons as the real robbers and evil monsters, heap their guilt for doing so on to them as scapegoats and then kill them to expiate their guilt. Dragons and the earth should be respected and lived with gratefully, sacrally and symbiotically. Dragons wreaking fiery vengeance on humans for stealing greedily from the earth can be read as a climate change allegory. More intense bush and forest fires in the age of global warming are the earth's vengeance for greedy mining and carboniferous capitalism (see Giblett, 2018b).

Notes

1 For a psychoanalytic ecological reading of mining, see Giblett (2011, chapter 9). For an environmentally friendly rewriting of the story of the dragon and Beowulf, see Giblett (2018b).
2 The 'birth of the dragon' for Ingersoll (2014, p. 13) took place in China. The dragon in traditional Chinese culture is an immense topic and outside the scope of this chapter and book and my expertise. For further (albeit dated and Orientalist) discussions, see Ingersoll (2014) and Hampden du Bose (2014). For a recent discussion of the dragon as an animal form in tai chi and other Taoist arts of health, see Fung Loy Kok Institute of Taoism (2008, pp. 72–74). The dragon is also present in Viking culture, again an immense topic and outside the scope of this chapter and book and my expertise. The dragon makes a repeated cameo appearance in Parker's book about the 'north men' in the form of Nithogg the Dragon in the legend of Yggdrasil and of Fafnir the Dragon in the legend of Sigurd, dragon tails and heads on the prow and stern of Viking longships, the longships themselves known as *snekke* ('serpent'), the earth-serpent spewing out venom, four corner posts of a bed with dragons carved on them, projecting gables of early Christian churches adorned with dragon-heads and dragons depicted on the side of the longships of Cnut's first fleet to England with fire pouring from their nostrils (Parker, 2015, pp. 43, 115, 119, 128, 130, 132, 135, 137, 248, 339, and 385 n93).
3 For two largely good and wise dragons, and for good and bad, wise and foolish humans, see Giblett (2018a and 2018b).
4 As we will see in chapter 8. Other instances are 'Bomber' Harris invoking *Hosea* 8: 7. See Giblett (2009, chapter 5).
5 For a heretical and environmentally friendly rewriting of the story of Saint George and the dragon, see Giblett (2018a and 2018b).
6 For a psychoanalytic ecological reading of marsh monsters and swamp serpents, see Giblett (1996, chapter 8).

References

Basile, G. 1992. The Merchant's Two Sons. *In*: J. Zipes, ed. *Spells of Enchantment: The Wondrous Fairy Tales of Western Culture*. New York: Penguin, pp. 39–47. First published 1634.

Beowulf and the Finnesburg Fragment 1950. J. Clark Hall, trans., new edition completely revised with an introduction by C. L. Wrenn and with prefatory remarks by J. R. R. Tolkien. London: George Allen and Unwin.

Beowulf. 1999. S. Heaney, trans. London: Faber and Faber.

Bose, H. du 2014. *The Dragon, Image, and Demon or, The Three Religions of China*. New York: Cosimo Classics. First published 1886.

Collins, M. 2012. *St George and the Dragon: The Making of English Identity*. London: CreateSpace.

Colonna, F. 1999. *Hypnerotomachia Poliphili: The Strife of Love in a Dream*. J. Godwin, trans. New York: Thames and Hudson. First published 1499.

Coover, R. 1991. The Dead Queen. Jack Zipes (ed.) *Spells of Enchantment: The Wondrous Fairy Tales of Western Culture*. New York: Viking Penguin, pp.704–711. First published 1973.

Dante Alighieri 1996. *The Inferno*. J. Ciardi, trans. New York: Random House. First published 1472.

Douglas, M. 1966. *Purity and Danger: An Analysis of the Concepts of Pollution and Taboo*. London: Routledge.

Fung Loy Kok Institute of Taoism. 2008. *A Path of Dual Cultivation: Teachings of the Fung Loy Kok Institute of Taoism*. Toronto: Fung Loy Kok Institute of Taoism.

Giblett, R. 1996. *Postmodern Wetlands: Culture, History, Ecology*. Edinburgh: Edinburgh University Press.

Giblett, R. 2009. *Landscapes of Culture and Nature*. Basingstoke: Palgrave Macmillan.

Giblett, R. 2018a. *The Dragon and Saint George: A Fairy Tale Novella*. Cambridge: Pegasus.

Giblett, R. 2018b. *Tales of Two Dragons*. London: Austin Macauley.

Greenblatt, S. 1982. Filthy Rites. *Daedalus*, 111(3), pp. 1–16.

Heaney, S. 2007. *Beowulf: A New Verse Translation*. Bilingual Edition. London: Faber and Faber.

Ingersoll, E. 2014. *Dragons and Dragon Lore*. New York: Cosimo Classics. First published 1928.

Malory, Sir T. 1969. *Le Morte D'Arthur: Volumes I and II*. Janet Cowen, ed. London: Penguin. First published 1485.

Malory, Sir T. 2001. *Le Morte D'Arthur: King Arthur and the Legends of the Round Table*. 'The Classic Rendition by Keith Baines'. New York: Signet Classics. First published 1962.

Parker, P. 2015. *The Northmen's Fury: A History of the Viking World*. London: Vintage.

Serle, G. 1971. *The Rush to be Rich: A History of the Colony of Victoria, 1883–1889*. Carlton: Melbourne University Press.

Stockton, F. 1992. The Griffin and the Minor Canon. *In:* J. Zipes, ed., *Spells of Enchantment: The Wondrous Fairy Tales of Western Culture*. New York: Penguin, pp. 462–473. First published 1885.

Tolkien, J. R. R. 1950. Prefatory Remarks. *In: Beowulf and the Finnesburg Fragment*, J. Clark Hall, trans., new edition completely revised with an introduction by C. Wrenn, London: George Allen and Unwin, pp. ix–xliii.

Tolkien, J. R. R. 2006. Beowulf: The Monsters and the Critics. Chap. 1 in *The Monsters and the Critics and Other Essays*, C. Tolkien, ed. London: HarperCollins, pp. 5–48. First published 1936.

Tolkien, J. R. R. 2010. *The Legend of Sigurd and Gudrún*, C. Tolkien, ed. London: HarperCollins.

Tolkien, J. R. R. 2013. *The Hobbit: Or, There and Back Again*. London: HarperCollins. First published 1937.

Tolkien, J. R. R. 2014. *Beowulf: A Translation and Commentary*. C. Tolkien, ed. London: HarperCollins.

Westermann, C. 1984. *Genesis 1–11: A Commentary*. J. Scullion, trans. London: SPCK.

4 Theology of watery monsters

Leviathan and crocodiles

Encounters between humans and crocodiles unfortunately occur in Australia with monotonous regularity and often with disastrous consequences for both parties, but also with cultural implications for how humans and animals cohabit planet Earth. This is especially the case for the way in which crocodiles are portrayed – usually as some sort of monster, including in the Bible and in the wider culture. For instance, in north Queensland in October 2004 a crocodile attacked a man in a tent and dragged him out of it. The man was only saved from a worse fate by the valiant efforts of a grandmother, who jumped on the back of the croc, which released the man and then proceeded to attack her. The print media headlined the story and captioned the accompanying photo with 'Gran who beat off croc attack' (2004). This headline gave a curiously Australian, and horrifyingly real, inflection to the immortal lines of a Tony Joe White song, 'Polk Salad Annie,/ 'gator's got your Grannie,/ chomp, chomp' (White, 1997). The newspaper story described the crocodile as 'a bloodthirsty predator'. This newspaper description of the crocodile was only jumping on the same bandwagon as its cousins in television news: they had already referred to the crocodile earlier in the week as 'the four-metre monster'. In both constructions of the crocodile as bloodthirsty predator and monster not only was its size, but also its use of its jaws and teeth as a potentially lethal weapon and the fearful possibility of being eaten were placed on the menu for the delectation of the media consumer who, over breakfast or dinner, could savour with relief that they were safe from being eaten.

This event and its media aftermath harks back to other incidents and representations involving crocodiles, such as Baby Bob and Steve Irwin, especially as there was a baby involved in the most recent incident. Fears for the safety of Baby Bob and the vilification of Steve Irwin from Indianapolis to Indooroopilly highlight human's fascination with, and fear of, crocodiles and their cousins, alligators. Such instances highlight the visceral nature of human bodily being and experience, particularly when it comes to food and feeding. With Steve holding a chicken in one hand to feed the croc and Bob in the other *not* to feed it, but with the possibility that it could feed on him, the boundary between being food and not food was evident for all to see, especially as baby and chicken were about the same size and colour. This raised fears of an oral kind, about who gets to eat and who gets to be eaten.

It also raised again the horrifying possibility that a croc might take a baby, just as a dingo had done in the Azaria Chamberlain case in 1980. And just as she was taken from a tent, so in the most recent case the baby was in a tent. As in the Lindy Chamberlain case, the Steve Irwin case also brought about the equally terrifying actuality of another moral panic and trial by media, with Steve accused of careless parenting like Michael Jackson, or attempted son-icide with the croc as weapon, or accomplice, or both. The portrayal of the croc as some sort of orally sadistic monster made it well-suited, if not typecast, to play this role which it has been playing for a long time – but only since humans have appeared on the earth long after it.

In this chapter I argue that the typecasting of the alligator and the crocodile as orally sadistic monsters is a projection of human desires and fears on to these non-human beings. These desires and fears of an oral nature are tied up with what Freud calls 'the uncanny'. The concept/metaphor of the uncanny is arguably Freud's greatest and most fruitful contribution to the study of culture, and to what I call, following Alexander Wilson, the cultures of natures (see Giblett, 2011, chapter 1). Yet the uncanny is interestingly and symptomatically absent from the classic dictionary of psychoanalytic terms (Laplanche, and Pontalis, 1973), and from a relatively recent account of the development and context of Freud's thought (Makari, 2008).

Roudinesco's intellectual and cultural biography of Freud devotes only half a page out of its 500 pages and a footnote in a chapter on 'Lumière sombres' ('Dark Enlightenment') to a discussion of the uncanny with the German *unheimlich* translated into French as 'l'inquiétante étrangeté' and as the 'l'inquiétante familier' following previous translations of Freud's essay into French (Roudinesco, 2014, pp. 269, 269 n.2 and 544). The French terms in cognate English terms of 'the strange disquietude' and 'the disquietingly familiar' are translated as 'the uncanny' in Roudinesco (2016, pp. 215, 468 n.2 and 548).

The uncanny fares even worse in Joel Whitebook's intellectual biography of Freud in which it is not discussed and Freud's essay on it is not even listed in the bibliography (Whitebook, 2017). The uncanny tends to be repressed in the secondary psychoanalytic literature. By contrast, the uncanny receives a full-length book study in cultural and literary studies (see Royle, 2003). It is also invoked in ecocultural studies, especially for the wetland as the uncanny place *par excellence* (see Giblett, 1996, especially chapter 2) and for the underside of the city (see Giblett, 2009, chapter 3; 2016).

For Freud, the alligator and the crocodile portrayed as orally sadistic monsters are vehicles and vectors of the uncanny. This combination gives rise to what could be called 'the monstrous uncanny' in which the fascinating and horrific are projected onto, and embodied in, an orally sadistic monster. The uncanny, I have argued elsewhere, counters the aesthetics of the sublime, the picturesque (pleasing prospects) and the beautiful (see Giblett, 1996, chapter 2 and 2011, chapter 3). Whereas the latter three privilege the distancing sense of sight, the uncanny engages the sense of smell which is much more up close and personal, though not as immediate as touch and taste. Smell is often commented upon in encounters, real or imagined, with an alligator or crocodile.

The monstrous uncanny, however, not only engages the olfactory but also the oral and tactile. The uncanny associated with smell (an association Freud made) is a sublimation to some extent of the uncanny associated with taste and touch. The uncanny is evoked by what is not seen, which could not only be what is smelt, but also what is tasted and touched, both of which are involved when being eaten. The fear of being killed and eaten, and the desire not to be, are understandable, and certainly characterize the human side of the relationship with alligators and crocodiles. It also presumably characterizes the other side too. The relationship between eater and eaten is ultimately non-reciprocal: only one being gets to eat and the other to be eaten. When humans venture into the habitat of alligator or crocodile they can be prey and 'being prey' as Val Plumwood describes it is a terrifying experience, but also instructive (as it was for her, as we shall see). Being prey highlights the non-reciprocal nature of the relationship. Besides consuming alligators and crocodiles as meat, humans also consume their habitat by destroying wetlands in a colonizing and non-reciprocal relationship.

The monstrous uncanny is also associated with the colonial unconscious, whether it be with William Bartram's and John Muir's encounters with an alligator in a Florida swamp, or Val Plumwood's and Sigmund Freud's accounts of stories about crocodiles in a New Guinea swamp. The return to the repressed involves a return not only to the individual's own repressed, but also to the culture's repressed. Both of these are figured in all these stories in association with the alligator or crocodile as an orally sadistic monster and the swamp as a grotesque place. Rather than reproducing this figuration of monstrosity, this chapter concludes by arguing for a relationship with alligators and crocodiles characterized by mutuality in which they and their habitat are respected and conserved. Animals, as Whatmore (2002, p. 32) advises, are 'best considered as strange persons, rather than familiar or exotic things'. In other words, they are best considered as what Haraway (2008) calls companion species, or as uncanny beings, rather than as monsters.

Freud and the smelly uncanny

Alligators and crocodiles have been living on the earth for 200 million years, barely half as long as sharks, but much longer than any other currently surviving, similar-sized genus of the animal kingdom. They are truly a 'blast from the past'. As such, they are vehicles and vectors for the uncanny. Freud (1985, p. 340) defined the uncanny as 'that class of the frightening which leads back to what is known of old and long familiar'. The uncanny is not only a return to the past, but also in quasi-Freudian terms 'a return *to* the repressed', including the colonial repressed (as I have argued elsewhere; see Giblett, 1996). What was repressed for the nineteenth-century, petit bourgeois and Viennese Freud was invariably sexual in nature.

More generally in patriarchal culture the repressed is what is corporeal, visceral, maternal and monstrous. These can include the sexual, oral and anal, all of which can be, and were, associated with the colonial. In Freud's 'The Uncanny' the crocodile emerges as a figure for the British colonial repressed to which he returns via the vehicle and vector of the artefacts of colonialism which bear the traces of other, alien or exotic places and peoples. The repressed does not have a fixed content or function but changes historically and varies culturally, though the crocodile and swamp as sites of the colonial unconscious have been with us for some time, as we will see. Crocodiles for the ancient Egyptians were sacred, as they are for Australian Aboriginal peoples.

How the sacred becomes the monstrous, and even demonic, is a vexed topic. For David Quammen in a reading of the biblical Leviathan in *Job* 41 the monster is both created by God and subdued by Him in order to instil humility in His people (Quammen, 2004, p. 13). The monstrous and divine have been split off from each other and the divine constitutes itself and its hold over its believers by claiming to have created the monster, and, in turn, by subduing it. This splitting occurs in and as the moment of the institution of history. In prehistory the monstrous and divine are coterminous and mutually defining, and the uncanny is the name of this imbrication. Indeed, the monstrously sacred/demonic is perhaps an apt definition of the uncanny. Just as the sublime functions as a secular theology in which the sublime stands in for God in a culture for which God is dead (see Giblett, 1996, 2011), so the uncanny operates as a kind of secular demonology with the alligator and crocodile as Satan and the swamp as hell (as we have seen and will see again below).

According to *The New Bible Dictionary* (Douglas, 1970, p. 729), Leviathan is usually considered to be 'some form of aquatic monster'. In *Psalm* 44: 26 'it is clearly of the sea and is generally thought to be the whale', though in the passage from *Job* 41 'most scholars agree that the creature is a crocodile'. The previous chapter of *Job* (40: 15–22) describes another monster translated as 'Behemoth' in the Authorised (or King James) Version. In the Revised Standard Version 'Behemoth' is footnoted as 'the hippopotamus'. The same version footnotes 'Leviathan' in the following chapter as 'the crocodile'. *The New Bible Dictionary* (Douglas, 1970, p. 138), concurs that 'Behemoth' in *Job* is 'the hippopotamus'. Yet the *New English Bible* (cited by Kelly, 2006, pp. 67 and 246) has 'crocodile' for 'Behemoth' in *Job* 40: 15 (this translation has 'whale' for 'Leviathan' in *Job* 41: 1). In this translation of the Bible the crocodile is both 'the chief of beasts' and 'the chief of God's beasts' (Kelly, 2006, p. 67). The crocodile is the king of the wet jungle, of the tropical wetland, the first, and best, work of God on the second day of creation (as we saw in the first chapter).

Dragons, by contrast, may be the worst work of God, or at least the worst creation coming out of primeval chaos. In *Psalm* 74: 13–14 Leviathan is associated with 'the dragons of the waters'. Perhaps Parker (2015, p. 132) had this verse in mind when he described the most advanced Viking longships as leviathans. Their longships were not only dragon-prowed, but dragons themselves. Dragons can be aquatic creatures, but they can also be aerial, infernal and terrestrial creatures (as we saw in the previous chapter). In *Psalm* 74 Trudinger (2001, p. 35) comments that 'great importance is given to Earth [. . .]; as cosmos, it is a place of harmony and order [. . .] The water-beings do not enjoy such a happy status. They are portrayed as God's opponents, suffer horrible injury and a degrading end'. Dragons and crocodiles are no exception. God is a dryland God, not a wetland God. He is a God of dryland agriculture and pastoralism. He is opposed to the wetland, its beings, and its aquaculture.

Leviathan has a much more ambiguous role and function in the Old Testament as a chaos monster and as a playmate than in the New Testament as the Devil or Satan (see Beal, 2002, p. 80). Whereas the dragon or Leviathan of the Old Testament is not associated with the Devil or Satan, in the last book of the New Testament the dragon or Leviathan is identified with the Devil or Satan (*Revelation* 9: 17 and 19 – as we saw in the previous chapter). In the Old Testament, as Beal (2002, p. 81) comments, 'neither Leviathan nor the sea monster nor any other monster is ever identified with the Devil or Satan'. In the last book of the New Testament, however, the dragon or Leviathan are identified with the Devil or Satan, and they are locked in the battle of 'good versus evil, God versus Devil' (Beal, 2002, p. 81).

Perhaps no animal has been more deified/demonized than the alligator and crocodile, the 'monarch of the marsh' and the 'king of beasts' of the tropical swamp. The alligatorian and the crocodilian have been repressed for a long time, at least since Freud's time, and it still persists. For Vollmar (1972, p. ix) 'crocodiles, alligators and caimans both horrify and fascinate'. In Freud's (1985, p. 339) terms, they are uncanny as he defined the uncanny as 'what is frightening – what arouses dread and horror' and I have defined the uncanny as what is both horrifying and fascinating (Giblett, 1996). Vollmar (1972, p.ix) suggests that 'lurid travellers' tales of evil reptiles lying loglike in tropical mud, ready to snatch and devour the unwary human, linger in the memory'. Freud (1985, p. 367; 2003, p. 151) developed the uncanny from reading one such tale, L. G. Moberly's 'Inexplicable', published in *The Strand Magazine* in 1917. This story harks back to the ur-travellers' tale of alligators: William Bartram's account of his travels, and encounter with alligators in a Florida swamp, first published in 1791 (I return to both these stories shortly).

A century and a half later in the mid-nineteenth century John Muir walked a thousand miles from Kentucky to the Gulf of Mexico through a Florida swamp where he encountered an alligator. Fifty years later and a year before Moberly's story, Muir's account of his encounter with an alligator, like Bartram's in a Florida swamp, was published posthumously. His reflections on this encounter are seen by some as a crucial transition in his thinking to a biocentric ethic. More recently, Val Plumwood's (1996, 1999, 2000a, b) gripping (perhaps the wrong word) autobiographical story of being crocodile prey published in (among other places) a collection by Travelers' Tales tries to avoid the lurid, but without success as I will argue later. For Plumwood, as with Muir, her experience was the occasion to reflect on her place in nature. If one is an uninvited guest in, or intruder into, the home of the alligator or crocodile, one can only expect to be prey as they are predators, or perhaps more precisely 'opportunist hunters and gorge feeders' (Kelly, 2006, p. 38). In other words, they are greedy feeders, not orally sadistic monsters.

In his reading of Moberly's story Freud downplays the role of real alligators and ignores the swamp as the place *par excellence* of the uncanny. If I were Freud,

I would psychoanalyse some long lost and repressed memories to do with his father, identify the crocodile as some sort of phallic symbol in Freud's lexicon of symbols and conclude that Freud's reading has something to do with his mother as the swamp is a maternal place.. Yet rather than psychoanalysing Freud's psychopathology, I want to analyse the psychogeopathology that portrays the alligator and the crocodile as an orally sadistic monster, to engage in the talking cure of a psychoanalytic ecology that would regard them and the swamp in less demonic and more sacral terms, and to promote ecomental health that would mean that these psychogeopathological symptoms did not arise in the first place (see Giblett, 1996, 2011).

Freud, perhaps in typical fashion, gives a three or four sentence summary of the story he 'came across' in a magazine. He does not give a reference other than saying that it appeared in 'a number of the English *Strand Magazine*' (Freud, 2003, p. 151). Some time ago I 'came across' two precise references to the story in Royle's *The Uncanny*, one to the original publication in *The Strand Magazine* and the other to a reprint in an anthology of stories from *The Strand Magazine* (Royle, 2003, pp. 140–141, n.3; Moberly, 1917 and 1991). Rather than concentrating on the slips and mistakes in Freud's retelling as he himself would do, I want to consider the gaps and absences, the symptomatic lacunae, of Freud's reading in order to reinstate alligators and crocodiles living in a swamp as a vehicle and vector of the uncanny and disinvest their construction as orally sadistic monsters.

In 'The Uncanny' Freud (2003, p. 151) relates how:

> During the isolation of the Great War, I came across a number of the English *Strand Magazine.* In it, among a number of pointless contributions, I read a story about a young couple who move into a furnished flat in which there is a curiously shaped table with crocodiles carved in the wood. Towards evening the flat is regularly pervaded by an unbearable and highly characteristic smell, and in the dark the tenants stumble over things and fancy they see something undefinable gliding over the stairs. In short, one is led to surmise that, owing to the presence of this table, the house is haunted by ghostly crocodiles, or that the wooden monsters come to life in the dark, or something of the sort. It was quite a naïve story, but its effect was extraordinarily uncanny.

Freud makes at least two mistakes in summarizing the story, the first of which is that the young couple move into a house, not a flat, furnished only with the table ('the table goes with the house' says the agent) and the table has carvings of alligators, not crocodiles, though Moberly's story itself slips between them as crocodiles, not alligators, inhabit New Guinea which is referenced in the story. These slips may be symptomatic of something more profound and if I were Freud I would no doubt think so and would analyse them (and him) for it. I am more interested, though, in the presences and absences in Freud's account, what he reproduces correctly and what he misses out altogether. Freud is not alone in referring to alligators or crocodiles as monsters as we have heard in a recent television report and as we shall see. There is a textual warrant for doing so in that the first-person narrator of the story refers to alligators as 'loathly monsters'. Yet this is not in relation to the carved, wooden alligators on the table, but in relation to a story within the story, a traveller's tale, about real, living alligators in a swamp in New Guinea that Freud does not consider at all as if he only read half the story by Moberly (more of that shortly).

First, the table. The first-person narrator (May, the wife of Freud's 'young couple') describes the effect the carved alligators had on her:

> as the light fell on the scaly bodies they had an extraordinary look of life, and the little sinister heads with the small evil eyes almost seemed to move. I shuddered and drew away from the table.
>
> (Moberly, 1917, p. 573 and 1991, p. 184)

May feels quite faint for a moment and says to the agent ''there is such a queer smell in here', [. . .] becoming all at once conscious of a strange and penetrating odour I had not before noticed'. She begins to suggest that it might be the drains but the agent is quick to interrupt and advise that they were ''set in order before the last tenant vacated the house [. . .] I have the sanitary people's certificate about them'' (Moberly, 1917, p. 573 and 1991, p. 185). This terrace house is, however, not a working-class 'swamp' whose bad air was the supposed vector of malaria in the miasmatic theory of disease and the target of the Sanitary Movement (see Giblett, 1996 and its references). The carved figures on the table are creatures of the swamp and smell is a vector of the uncanny in the psychoanalytic theory of the psychopathology of everyday life in modernity (see Giblett, 1996).

May's husband Hugh's first encounter with the table is just as uncanny as hers. He runs his fingers over the carved surface of the table and rests them on the head of one of the alligators, 'a head fashioned with such skill that its loathsome naturalness made one shudder. "Good heavens, May, the things look so lifelike I could almost have sworn one of them squirmed"' (Moberly, 1917, p. 574 and 1991, p. 186). Inanimate things coming to life, or seeming to do so, or imitating life, is for Freud one of the features and vectors of the uncanny.

The figuration of the alligator as 'a thing', 'this thing' and 'the thing' persists into recent fiction as this is precisely how T. Coraghessan Boyle figures one in his 1990 novel *East is east* (Boyle, 1990, pp. 335–336). 'The Thing' has been defined by Kristeva (1989, p. 13) as:

> the real that does not lend itself to signification, the centre of attraction and repulsion, the seat of sexuality from which the object of desire will become separated [. . .] the Thing is an imagined sun, bright and black at the same time.

'The Thing' occupies a zone prior to and outside signification, what Kristeva (1984) elsewhere calls 'signifiance', which is also prior to and outside the separation of subject and subject in what Kristeva (1982) also calls abject. It is a creature of the uncanny, the centre and site of fascination and horror, the vehicle and vector of the monstrous.

Hugh and May are visited by a friend called Jack Wilding for whom the carved alligators trigger memories not related by Freud in his recounting of the story, which highlights the importance of the fact that the table is located in a house, not a flat. After dinner 'on a delicious May night' the two men friends are chatting with the windows open. The spring smells of the garden waft into the room:

> when all at once the drifting sweetness from without was tainted by that same strange odour which we had noticed once or twice before [. . .] As it drifted across the room our guest suddenly sat bolt upright in his chair, and a curious greyness overspread his naturally bronzed complexion. 'My God!' he said, what is that? And why does it smell the same – the same – ' His sentence trailed off into silence, and in the intense stillness following his strange words I heard a sound which, for some reason I could not pretend to explain, gave me a feeling of cold fear. I can only describe the sound as like a far-away bellowing – not precisely the bellowing of cattle, but a more sinister, more horrible sound, pregnant with evil. 'You hear it too?' Jack Wilding questioned, under his breath. . . 'And the stench is here too! Good God! If I thought I should ever have to cross that swamp again I would go mad.'
>
> (Moberly, 1917, p. 576 and 1991, p. 190)

The immediacy of the sense of smell takes Jack back to what was old and long familiar, and to what was long forgotten, if not repressed. Jack then recovers himself sufficiently to say:

> 'I must have had a nightmare – a waking nightmare,' he said, looking around him. 'I could have sworn that I smelt the alligator swamp in New Guinea, the place where – ' He broke off short. 'I heard the loathsome brutes bellowing,' he began again; 'but, of course – of course, it was merely some association of ideas,'
>
> (Moberly, 1917, p. 577 and 1991, p. 190)

Hugh indicates the table and suggests that it was the trigger for Jack's association of ideas:

> Jack turned and glanced at the table, and he recoiled when he saw the grinning heads lying amongst the crusted delicacy of leaves and flowers. 'Loathsome beasts!' he said, and again his voice shook [. . .] I crossed an alligator swamp once with a friend [. . .] It was dark, the place swarmed with those unspeakable devils, their stench was everywhere. It was dark – and poor old Danson' – he paused, as if speech were almost impossible – 'they dragged him off the path of the logs in the darkness' [. . .] Somehow his words brought before me the hideous swamp, the darkness, the loathly monsters waiting for their prey, and the remembrance of just such an incident in a book I had once read flashed into my mind.
>
> (Moberly, 1917, p. 578 and 1991, p. 191)

The original *Strand Magazine* publication of the story has accompanying illustrations by Dudley Tennant, one of which is of this story within the story, complete with a glimpse of both the table and the log path, and a depiction of Jack as both the teller of the tale and character in it (Moberly, 1917, p. 577).

Jack goes on to relate how '"the place swarmed with those unspeakable devils."' Swarming creatures, in biblical terms, are 'an abomination'. They are neither fish nor flesh nor fowl. They neither just swim nor walk nor fly, but do all three. These alligators are no exception. When the wooden alligators come to life they always seem to be sliding or slithering between Jack's or Hugh's feet (but not May's, as presumably she keeps her legs together like a lady). The housekeeper later describes how they go 'slithering' and 'running on their underneaths' (Moberly, 1917, p. 574 and 1991, p. 193). They do not walk on all fours like domesticated animals. They do not separate their grotesque lower bodily stratum from the grotesque lower earthly stratum, but are part and parcel of it. The abominable is also almost beyond words. It is inexplicable as the story concludes and as its title signals.

Muir and the terror-stricken mouthful

A year earlier than the publication of Moberly's story, John Muir's *A Thousand Mile Walk to the Gulf* was published posthumously, relating a journey he had undertaken almost fifty years earlier. In 1867 Muir, 'one of the seminal figures in the history of modern environmental thought' according to Frederick Turner in his introduction to the Penguin Nature Library edition of Muir's book, encountered an alligator 'on the margin of a stagnant pool' in a Florida swamp which prompted Muir to reflect that:

> These independent inhabitants of the sluggish waters of this low coast cannot be called the friends of man [. . .] Many good people believe that alligators were created by the Devil, thus accounting for their all-consuming appetite and ugliness. But doubtless these creatures are happy and fill the place assigned them by the great Creator of us all. Fierce and cruel they appear to us, but beautiful in the eyes of God. They, also, are his children, for he hears their cries, cares for them tenderly, and provides their daily bread. The antipathies existing in the Lord's great animal family must be wisely planned, like balanced repulsion and attraction in the mineral kingdom. How narrow we selfish, conceited creatures are in our sympathies! How blind to the rights of all the rest of creation! With what dismal irreverence we speak of our fellow mortals! Though alligators, snakes, etc. naturally repel us, they are not mysterious evils. They dwell happily in these flowery wilds, are part of God's family, unfallen, undepraved, and cared for with the same species of tenderness and love as is bestowed on angels in heaven or saints on earth. I think that most of the antipathies which haunt and terrify us are morbid productions of ignorance and weakness. I have better thoughts of those alligators now that I have seen them at home. Honorable representatives of the great saurians of

> an older creation, may you long enjoy your lilies and rushes, and be blessed now and then with a mouthful of terror-stricken man by way of dainty!
>
> (Muir, 1992, pp. 98–9)

Here endeth the sermon on 'how I learned to stopped hating alligators, and learnt to love them'. Only a true greenie would hug an alligator metaphorically as readily as they would a tree in actuality. Perhaps needless to say, Muir was not a mouthful of terror-stricken humanity for the alligators of a Florida swamp. He may have thought differently and not have been quite so jaunty if he had been as Val Pumwood was, and did, as will see.

Bartram and the greedy monster

The story of the abominable crocodile and the monstrous alligator lurking in the uncanny swamp surfaces much earlier in modern Western culture than with Freud, Muir and Moberly in the early twentieth century. William Bartram in the late eighteenth century travelled to a Florida swamp where he encountered 'the subtle, greedy alligator' about to devour 'the voracious trout', the eater eaten, the preyer predated by a larger predator:

> His enormous body swells. His plaited tail brandished high floats upon the lake. The waters like a cataract descend from his opening jaws. Clouds of smoke issue from his dilated nostrils. The earth trembles with his thunder.
>
> (Bartram, 1998, p. 75)

Two hundred years later Glasgow takes this last sentence as the sub-title for his social history of the American alligator and suggests that Bartram 'certainly wins, hands down, any contest for colourful alligator writing' (Glasgow, 1991, p. 31). There is certainly plenty of that in the intervening two centuries, as Glasgow's study shows. Bartram's description is a lurid traveller's tale complete with exaggeration that the vapour is smoke, for which he has a biblical warrant (*Job* 41: 20), though in a later encounter he describes how 'the vapour ascends from his nostrils like smoke' (Bartram, 1998, p. 82). It would only take the addition of fire from the alligator's mouth and wings on its back to complete the construction of the alligator as a dragon. Bartram's alligators are, as Slaughter (1996, p. 200) points out, 'monsters on a heroic scale. Bartram's drawing, *The Alegator* [sic] of *St Johns* [now housed in the Natural History Museum in London], resembles medieval representations of the dragons battled by St George'. Bartram on a couple of occasions refers to 'the monster' (Bartram, 1998, pp. 77, 80). Bartram is a latter-day St George to the alligator's dragon (with some considerable biblical and cultural warrant for making this association (as we have seen previously)).

If, as I have argued elsewhere (see Giblett, 1996), the swamp is a secular underworld into which the hero of the modern adventure romance has to descend and in which he has to overcome monsters, then Bartram's travels involve him not only descending into Florida swamps but also overcoming their resident monsters, alligators – as he does later to emerge 'victorious, or at least [having] made a safe retreat' (Bartram, 1998, p. 80) as he concedes. Bartram leaves us in no doubt that we are in the underworld when he describes 'the boiling surface of the lake' (p. 75) created by two alligators engaged in 'horrid combat' (p. 76). The alligator swamp is a hellish, hot place and a place where the elements of water (lagoon) and fire (or heat) mix, as do the elements of air, earth and water created by the alligator itself. The 'dreadful roar' (p. 76) or the 'terrifying roar' of the alligator 'resembles very heavy distant thunder, not only shaking the air and waters, but causing the earth to tremble' (p. 82).

The alligator mixes the elements of earth, air, water and fire (thunder and lightning are the fire in the sky) just as the wetland does more generally (as we saw in chapter 1; see also Giblett, 1996). Instead of these elements staying put in their proper place, the alligator and the wetland mix them up and violate the order of things that assigns them to a fixed and stable category. Even the smoke/vapour exhaling from the alligator's nostrils upsets the distinction between air, water and fire (where there is smoke, there's fire). The fact that the alligator in the water 'resembles, at a distance, a great chunk of wood floating about' (Bartram, 1998, p. 82) means that it pretends to be solid and earthy when in fact it is monstrous and slimy.

And orally sadistic to boot. If there is one feature of the alligator that more than any other arouses dread and horror for Bartram it is the fear of being 'dragged out of the boat and instantly devoured' (Bartram, 1998, p. 76) by the greedy monster. Its jaws are prodigious, with the upper jaw opening at right angles to the lower one according to Bartram (p. 82) but mistakenly for Glasgow (1991, p. 31). Bartram also found that they are capable of 'belching floods of water' (Bartram, 1998, p. 76). The jaws of the alligator emit terrifying roars and belch water, and so give out noise and liquid. They also take in liquid in the form of water and blood, and solid in the form of flesh, animal or human. Their jaws give and take. Their jaws are a transportational zone between the inside and outside of the body through which matter passes.

Alligators in a Florida swamp function in both Bartram's and Muir's texts as figures for the British colonial and American cultural unconscious. Florida and the American South more generally with its swamps have been repressed in the collective psyche of the American North and have functioned as a figure of the primitive and backward (see Giblett, 1996 and references). Relatedly, American wetlands have been drained and their creatures commodified from early colonial times. In Bartram's case he is returned to this repressed only to try to continue to repress it; in Muir's, he embraces it – albeit metaphorically.

Plumwood and great-toothed jaws

Unlike both Muir and Bartram and their encounters with an alligator in a Florida swamp, but like Danson in Moberly's story, the Australian ecofeminist and ecophilosopher, Val Plumwood, was a mouthful of terror-stricken humanity for a crocodile in Kakadu National Park in northern Australia in 1985. She lived to tell the tale, though, unlike Danson, but it took some time for her to be able to tell it. I am not going to attempt to retell the story in all its details and if I attempted to sum it up by saying it was a gripping and gut-wrenching story I would be making bad puns. Plumwood's story is worth reading for itself. By reading it critically I am not belittling or demeaning her experience, which was traumatic to state the obvious, nor her individually as she was heroic to say the least. Rather, as with Freud, I am trying to critique the patriarchal and colonial elements that emerge in it despite her best efforts to keep them at bay.

Her story has some explicit elements of the uncanny, such as her description of 'the unfamiliar sensation of being watched' and her 'whispering sense of unease' prompted, not by the sight of a crocodile, but of 'a strange rock formation' (Plumwood, 1996, pp. 33–4; 1999, p. 78; 2000a, p. 57; 2000b, p. 130) which is a portent of what is to come. Plumwood hears whispers of unease and observes the strange rock formation so she is in a heightened state of sensory alertness. When the crocodile attacks her flimsy fibre-glass canoe 'the unheard of was happening' (Plumwood, 1996, p. 34; 1999, p. 78; 2000a, p. 131; 2000b, p. 57). The uncanny is evoked by hearing (or not) and smelling, by what can be heard or not, what can be smelt, but generally not by seeing. Deathly silence evokes the uncanny. In this case, being seen also evokes it. 'Being prey' as she calls it, is the result of being seen. To be prey one is eventually seen. The crocodile stalks its prey using the senses of hearing, smell and sight. Plumwood is not the subject of the gaze, but its object. She hears whispers of unease and hears (and sees) the unheard of, a crocodile attacking a canoe.

The ten-year gap between 'being prey' and publishing her account and reflections on this event under this title attests perhaps to the difficulty for her of coming to terms with her experience and expressing it. Part of the difficulty was not only the trauma of the attack itself but also the way in which her story was subjected to what she called 'the cultural drive to represent it [the attack] in terms of the masculinist monster myth: the master narrative' (Plumwood, 1996, p. 40; 1999, p. 85; 2000a, p. 139; 2000b, p. 59). In this myth, the crocodile is constructed as a ravening, orally sadistic monster who (or which) rapes and eats his (and it is always a male in the myth) innocent female victim. Yet Plumwood reproduces the myth herself when, for example, she earlier describes in lurid terms how she had 'a blurred, incredulous vision of great toothed jaws bursting from the water' that then 'seized [her] between the legs in a red-hot pincer grip' (Plumwood, 1996, p. 35; 1999, p. 79; 2000a, p. 131; 2000b, p. 57). The crocodile is figured as a monstrous, orally sadistic and reptilian cousin of 'Jaws', the monstrous shark.

Rather than only construing her experience in good ecological terms of being prey and so being a part of the food chain, Plumwood also turns in the longer version of her story to the mythology of New Guinea (and why not that of Australian Aborigines one wonders) as a way of accounting for her experience that does not merely deny or repress its mythological elements and significance. She suggests that 'crocodiles are masters of water' and goes on to argue that:

> the crocodile is an exploiter of the great planetary dualism of land and water. As Papua New Guinea writer Vincent Eri suggests in his novel, *The Crocodile*, the creature is a sort of magician: its technique is to steal the Other, the creature of the land, away into its world of water where it has complete mastery over it. Water is the key to the crocodile's power, and even large crocodiles rarely attack in its absence. The crocodile is then a boundary inhabitant.
> (Plumwood, 1996, p. 39; 1999, p. 84; 2000a, p. 137)

In other words, the crocodile, like the alligator, is a wetland inhabitant, an inhabitant of the intermediary zone between dry land and deep water that crosses 'the boundary' between land and water and upsets the dualism between them (see Giblett, 1996). Yet, like Freud who ignores the story within the story of the alligator swamp in New Guinea (and like Royle who ignores the alligators and the swamp in the story within the story), and like both who ignore these aspects as the vector and vehicle for the uncanny, Plumwood overlooks the fact that the wetland with its distinctive features is the crocodile's habitat. She describes the latter as 'the swamp' and contrasts that with 'Kakadu's wetlands' with their 'dream-like beauty' enticing her into 'a joyous afternoon's idyll' at the beginning of the story. She describes how here she 'glutted' herself on 'the magical beauty [. . .] of the lily lagoons untroubled by crocodiles' (Plumwood, 1996, p. 33; 1999, p. 76; 2000a, p. 128; 2000b, p. 56). She moralizes the pastoral world of the wetland as good and heavenly – good enough to eat with herself as metaphorical glutton – and the swamp as bad and hellish, the place of the orally sadistic and gluttonous crocodile. For her, the crocodile lives in a swamp whereas the tourist visits a wetland; in brief, swamp is bad, wetland is good.

Naturally, she would rather eat than be eaten (and wouldn't we all?) Just as she is being watched, but prefers to watch, so she prefers to 'eat' the beauty gluttonously than to be eaten greedily by the crocodile. Who can blame her? The point is, though, oral sadism rules, okay? She is a metaphorical glutton and the crocodile is figured as an orally sadistic monster. Tourist and crocodile are going about their normal, everyday business subscribing in both cases to the master narrative – despite Plumwood's best efforts to avoid it and do otherwise. The tourist watches (preferably without being watched) and consumes beauty gluttonously through his or her eyes without being consumed; the crocodile watches the tourist and sometimes consumes him or her through his or her mouth. Both are just doing their own thing in this mythology of modernity with its non-reciprocal and non-symbiotic positions of mastery of watching and eating, rather than being watched and eaten.

As the Kakadu wetland for Plumwood is good and heavenly, the crocodile swamp by implication is bad and hellish, though the wetland and the swamp are one and the same place – they are just figured differently. The Kakadu wetland is Eden before the Fall, or before the appearance of the serpent, here transformed into the crocodile. It is also the place of good magic, but crocodiles are excluded from this world of good magic as they bring trouble into paradise by enticing creatures of the land with bad magic into a watery grave, into a wet underworld, into its world – the swamp. The good magic is a sanitized white magic enticing Plumwood into the beautiful pastoral idyll of the wetland split off from, and valorized over, the black magic of the crocodile enticing her into the horrifying black water of the swamp. Magic is just magic, though, without the moralization. Plumwood (curiously for an ecofeminist) reproduces not only the patriarchal, Western moralization of the wetlandscape, but also its dualisms and spatial metaphysics and poetics of land and water, good and bad, white and black, heaven and hell, above and below (see Giblett, 1996).

Yet humans are not simply or exclusively creatures of the land, just as crocodiles are not simply or exclusively creatures of the water. Alligators and crocodiles are creatures who live in the two elements of earth and water, just as their wetland home mixes these elements (Strawn, 1997, p. 14; Giblett, 1996). Humans are also creatures of water as we are predominantly made up of water. Our beginnings as individuals are in the watery world of the womb and our beginnings as a species in evolutionary terms are in the womby world of water (as we saw in chapter 1; see also Giblett, 1996). Humans also have vestigial reptilian parts of the brain. Humans are meat for crocodiles (and vice versa), but we are also very distant cousins. We are both wetland creatures living on a complementary, non-dualistic planet of land and water, in short, of waterland.

Alligators and crocodiles in Haraway's (2008) terms are 'companion species' (though not in the same way as pets are) who occupy what she also calls a 'contact zone' between human beings and non-human beings (though again not in the same way as the human home, or domestic space, is for pets). The quaking zone is a contact zone between humans and non-humans (such as alligators and crocodiles in native quaking zones, and lice and rats in feral quaking zones of the trenches and of the ruined cities of world warfare and of urban slums). This contact zone can be phatic involving communication via the acts and senses of touching, smelling, hearing and seeing, or it can be phagic involving consumption via the acts of biting, chewing, etc. and the sense of taste. The phatic contact zone as a rule occurs with pets; the phagic with non-pets. If the pet becomes phagic, it ceases to be pet (and phatic) and may be 'put down'.

The uncanniness of alligators and crocodiles lies in the fact that they and modern urban humans live in the same earth home (ecosphere), but not in the same house. Modern urban humans may live in the earthly wetland home of alligators and crocodiles, but the latter do not live in human houses and gardens as a rule. Alligators and crocodiles are (un)homely creatures, like human beings, of the terraqueous globe as Serres (1997, p. xiii) calls the earth. For Haraway (2008, p. 45) 'the familiar is always where the uncanny lurks. Further, the uncanny is where value becomes flesh again'. The familial – the family of 'Man', the family of reptilians (including humans) – is where the alligatorian and crocodilian lurk just beneath the surface of the wetlands of the earth. Flesh is what carnivores – human, alligatorian and crocodilian – eat. The body becomes flesh in the uncanny; the value of the body is chomped into the flesh of a 'mouthful of terror-stricken humanity' as for Muir and Plumwood in the jaws of an alligator or crocodile, or of a mouthful of crocodile burger in my jaws as when I visited a crocodile farm outside Lusaka in Zambia in 2007. Eat or be eaten indeed.

Plumwood draws on Eri's novel to support her argument about crocodiles and water, but her reading has no real basis or textual warrant. Beside the perhaps customary references in the novel to 'the monstrous crocodile' and to 'the horrifying creature' (Eri, 1973, pp. 108, 113), the crocodile of the title does not figure (in two senses of the word) much in the novel. It is certainly not used, as Plumwood later suggests (1996, p. 39; 1999, p. 85; 2000a, p. 138), as:

> a metaphor for the relationship between colonized indigenous culture and colonizing Western culture. If the crocodile-magician-colonizer can drag you completely into its medium, you have little chance; if you can somehow manage to retain a hold on your medium, you may survive.

If the crocodile is used in Eri's novel as a metaphor for anything, it is as a device for explaining the inexplicable, unlike for Moberly for whom the crocodile *is* the inexplicable. When Mitoro, the wife of the central character Hoiri, disappears the whole village maintains that a crocodile took her. Interestingly in one Aboriginal story the crocodile is a wife-stealer too (see Mudrooroo, 1994, p. 33). The men of the village hunt down a crocodile and Hoiri kills it in an act of revenge that he is privileged to perform as the victim. Yet at the end of the novel Hoiri encounters Mitori. She does not acknowledge him and no explanation for her leaving him is given other than that she is under the power of the magicians who also control or transform themselves into crocodiles.

The crocodile is a scapegoat onto which her sin of leaving her husband is heaped (as she cannot be found and punished) and the village is expiated. The crocodile is a creature of the wetland figured as what Lyotard called a 'scapeland', not only in the sense of the anti- or counter-landscape but also in the sense of the sacrificial victim onto, or into, which the sins of the community are heaped and expiated (see Giblett, 1996). Perhaps in modern Western medical terms Mitoro was suffering from post-natal depression following the birth of their son and left Hoiri as a result. This would highlight the connection between melancholia and wetlands, and the creatures of the wetland (see Giblett, 1996). Alternatively, perhaps she simply ran away with another man. The crocodile-magician is a part of indigenous culture that is not necessarily good, or acts for good, but functions to explain the inexplicable, unlike in Moberly's story of this title.

The colonizer, on the other hand, is largely represented by piggish Patrol Officers or draconian District Officers. One of them is 'referred to as "the crocodile", a title that was one of praise rather than abuse' (Eri, 1973, p. 141), presumably because of his power, cruelty and ugliness. He is monstrous and horrifying, like the crocodile. The colonizer certainly wants to drag the indigene into his medium represented by patrols, prisons, cities and warships. The crocodile does not represent this world. It represents an unsettling liminal zone of indigenous culture that is not necessarily good or bad, but it is certainly not the colonizer's culture, or a metaphor for it other than for the monstrous, horrifying and inexplicable in it, and all cultures, as in Moberly's story.

Plumwood associates the crocodile with the relationship between colonizer and colonized; Freud associates the uncanny with an artefact of colonialism with carved crocodiles that seem to, or do, come alive. Both are associated with a New Guinea swamp. The crocodile surfaces uncannily here in an Australian ecofeminist text quoting a novel, a colonial genre, written in a colonized culture. It also emerges in Freud's Viennese cultural and psychoanalytic repressed via a British magazine as a vector for the uncanny. In both texts the crocodile surfaces as a figure for the British colonial unconscious – repressed, but returning in their slips, gaps and lacunae via the vehicle and vector of the artefacts of colonialism which bear the traces of other, alien or exotic places and peoples.

In Beth Yahp's 1992 novel, *The Crocodile Fury*, set in Malaysia the crocodile functions in a slightly different way as a projection (or 'symbol') of repressed desires and fears. For the first-person narrator's grandmother male sexual aggression is represented by 'the land crocodile', 'a creature who can't be controlled'. For her,

> this beast, this human-shaped terrorizer of innocent girls [. . .] lives on the edges of the jungle so he gets the best of both worlds [. . .] when the urge takes him, out he creeps to jump at victims [. . .] When he's on a rampage his whole body swells, he lets out a sound that is low and throbbing, that fills young girls' and old women's heads with the hum of a thousand insects, the shudder of earthquakes, the toppling of great jungle trees, so they fall over faint and panting. When the crocodile is on a rampage he gives off a red glow. When a young girl is deflowered or a man seduces some else's wife, or your daughter is stolen, your faithful assistant led astray, you can be sure it's with the help of the land crocodile. A girl unlucky enough to be touched by the crocodile will forever be branded with wild ideas, with tempestuous fevers. She'll never listen to her elders, she'll never do what she's told. Such a girl will become a lifetime victim of passions so itchy she'll never be able to sit still.
>
> (Yahp, 1992, p. 18)

Redolent of Bartram's alligator, the land crocodile is a creature of the quaking zone of earthquakes and of the liminal zone, not between land and water, but between city and jungle. He crosses the boundary between the two to get the best of both worlds. Like Eri's crocodile, the land crocodile is a convenient explanatory device, a heuristic trope, for explaining inexplicable, abnormal, or unusual, or typical male sexual behaviour.

Monstrosity to mutuality

My three encounters or interactions with alligators and crocodiles have been neither as spectacular nor as close as Plumwood's, and pretty tame by comparison, but instructive nevertheless along the lines I have been pursuing in this chapter about living together on earth with alligators and crocodiles as earthly companion species and sharing a bioregion with them. I visited the Kimberley in 1992 and ventured into salt-water crocodile country down the lower Ord River in a small boat operated by a tour guide. This venture was part of research for a book on Western Australian wetlands with Hugh Webb. The photographer accompanying us, Simon Neville, took photographs of crocodiles which involved getting within a few metres of the shore. Greg, the guide, pointed out that he only went this close when there was enough water under the boat so that if the croc decided it wanted to go into the water it could go under the boat, rather than launching itself across the top of the boat and taking one of us with him or her. We all appreciated this consideration of *not* being prey!

My other interactions with alligators and crocodiles were perhaps a bit tamer and much more mainstream touristic. My most recent interaction with crocodiles was in 2016 in Costa Rica on a guided tour of the famous Tarcoles River. The boat trip included a sighting of the biggest crocodile on the river called 'Mike Tyson'. I couldn't help wondering about Plumwood's story, though, and her best attempts to keep the masculinist monster myth and the master narrative at bay as they seemed to be re-emerging with Mike Tyson the boxer and the crocodile. The captain of the boat managed to herd Mike towards the shore, at which point the guide jumped out of the boat into the water and coaxed Mike towards the shallows by waving a yellow cloth. Mike came closer while the guide continued to wave the cloth. Mike obligingly opened his mouth and the guide got the camera of one of my fellow tourists and took a photo a few inches from Mike's nose. I asked the captain of the boat if the guide got paid danger money. Mike must have been well-fed, or unthreatened, or familiar with the guide, though he said that they hadn't see Mike for while and he had obligingly shown up for our visit.

My one and only interaction with alligators was perhaps a bit tamer and also much more mainstream touristic. Like Bartram and Muir, it was in a Florida swamp. Wakulla Springs is a popular picnic and swimming spot outside of Tallahassee. According to the inevitable brochure for the place, Wakulla is a Seminole Creek word that means 'strange and mysterious waters', in a word, uncanny, but not monstrous, like the alligators. Such films as *Creature from the Black Lagoon* and some of the old Tarzan films with Johnny Weissmuller were shot there. You get the general picture of swampy jungle. It is alligator habitat, too, so the alligators are supposed to stay on one side of the river and people on the other. A lifeguard is stationed on the people side and if an alligator ventures across to this side the lifeguard evidently yells out, 'Alligator on the people side!' Everyone is supposed to get out of the water. The people side is roped off to stop people going across to the alligator side. A short boat tour takes tourists downstream and into alligator territory for a closer look. I saw two alligators and two turtles lying on the same log. Perhaps this is an instance of mutuality with two species living together and sharing the same living space, just as alligators and humans share the swimming hole with a degree of respect on the part of humans for the alligators in this place, but not in many others (see Strawn, 1997, esp. p. 170; Giblett, 2011). Alligators, crocodiles and humans should be able to live together in a psycho-symbiotic relationship and pursue their livelihoods in the same bioregion on the same earth.

References

Bartram, W. 1998. *Travels: Naturalist's Edition*. F. Harper, ed. Athens, GA: University of Georgia Press. First published 1791.

Beal, T. 2002. *Religion and its Monsters*. New York: Routledge.

Boyle, T. 1990. *East is East*. New York: Penguin.

Douglas, J. ed. 1970. *New Bible Dictionary*. London: Inter-Varsity Press.

Eri, V. 1973. *The Crocodile*. Ringwood, Victoria: Penguin.

Freud, S. 1985. The 'Uncanny'. *In: Art and Literature*, J. Strachey, trans., Penguin Freud Library 14, Harmondsworth: Penguin, pp. 335–376. First published 1919.

Freud, S. 2003. *The Uncanny*. D. Mclintock, trans. London: Penguin. First published 1919.

Giblett, R. 1996. *Postmodern Wetlands: Culture, History, Ecology*. Edinburgh: Edinburgh University Press.

Giblett, R. 2009. *Landscapes of Culture and Nature*. Basingstoke: Palgrave Macmillan.

Giblett, R. 2011. *People and Places of Nature and Culture*. Bristol: Intellect Books.

Giblett, R. 2016. *Cities and Wetlands: The Return of the Repressed in Nature and Culture*. London: Bloomsbury.

Glasgow, V. 1991. *A Social History of the American Alligator: The Earth Trembles with his Thunder*. New York: St Martin's Press.

'Gran who Beat off Croc Attack' 2004. *The West Australian*, 16 Oct., p. 40.

Haraway, D. 2008. *When Species Meet*. Minneapolis: University of Minnesota Press.

Kelly, L. 2006. *Crocodile: Evolution's Greatest Survivor*. Crows Nest, NSW: Allen and Unwin.

Kristeva, J. 1982. *Powers of Horror: An Essay on Abjection*. L. Roudiez, trans. New York: Columbia University Press.

Kristeva, J. 1984. *Revolution in Poetic Language*. M. Waller, trans. New York: Columbia University Press.

Kristeva, J. 1989. *Black Sun: Depression and Melancholia*. L. Roudiez, trans. New York: Columbia University Press.

Laplanche, J. and Pontalis, J.-B. 1973. *The Language of Psychoanalysis*. Donald Nicholson-Smith, trans. London: Hogarth.

Makari, G. 2008. *Revolution in Mind: The Creation of Psychoanalysis*. New York: Harper.

Moberly, L. G. 1917. Inexplicable. *The Strand Magazine*, 54(324), pp. 572–581.

Moberly, L. G. 1991. Inexplicable. *In:* J. Adrian, ed., *Strange tales from The Strand*, Oxford: Oxford University Press, pp. 183–195.

Mudrooroo. 1994. Crocodiles. *In: Aboriginal Mythology: An A-Z Spanning the History of Aboriginal Mythology from the Earliest Legends to the Present Day*, London, HarperCollins, pp. 33–35.

Muir, J. 1992. *A Thousand-Mile Walk to the Gulf*. New York: Penguin. First published 1916.

Parker, P. 2015. *The Northmen's Fury: A History of the Viking World*. London: Vintage.

Plumwood, V. 1996. Being Prey, *Terra Nova*, 1(3), pp. 32–44.

Plumwood, V. 1999. Being Prey. *In*: D. Rothenberg and M. Ulvaus, eds, *The New Earth Reader: The Best of Terra Nova*, Cambridge, MA: MIT Press, pp. 76–91.

Plumwood, V. 2000a. Being prey. *In* J. O'Reilly, S. O'Reilly and R. Sterling, eds, *The Ultimate Journey: Inspiring Stories of Living and Dying*, San Francisco: Travelers' Tales, pp. 128–146.
Plumwood, V. 2000b. Being prey. *UTNE reader*, July-August, pp. 56–61.
Quammen, D. 2004. *Monster of God: The Man-Eating Predator in the Jungles of History and the Mind*. London: Hutchinson.
Roudinesco, É. 2014. *Sigmund Freud en son Temps et dans le Nôtre*. Paris: Seuil.
Roudinesco, É. 2016. *Freud: In his Time and Ours*. C. Porter, trans. Cambridge, MA: Harvard University Press.
Royle, N. 2003. *The Uncanny*. New York: Routledge.
Serres, M. 1997. *The Troubadour of Knowledge*. S. Glaser with W. Paulson, trans. Ann Arbor: University of Michigan Press.
Slaughter, T. 1996. *The Natures of John and William Bartram*. New York: Random House.
Strawn, M. 1997. *Alligators, Prehistoric Presence in the American Landscape*, Baltimore: The Johns Hopkins University Press.
Trudinger, P. 2001. Friend or Foe? Earth, Sea and *Chaoskampf* in the *Psalms*. *In*: N. Habel, ed., *The Earth Story in the Psalms and the Prophets*, Sheffield: Sheffield Academic Press, pp. 29–41.
Vollmar, F. 1972. Preface. *In*: C. Guggisberg, *Crocodiles: Their Natural History, Folklore and Conservation*, Mount Eliza, Victoria: Wren, pp. ix–x.
Whatmore, S. 2002. *Hybrid Geographies: Natures Cultures Spaces*. London: Sage.
White, T. 1997. Polk Salad Annie. *Tony Joe White Collection* [CD]. Festival Records.
Whitebook, J. 2017. *Freud: An Intellectual Biography*. Cambridge: Cambridge University Press.
Yahp, B. 1992. *The Crocodile Fury*. Sydney: Angus and Robertson.

5 'The Earth is the Lord's, and its inhabitants'

The Psalmists' environmental theologies

The biblical book of *Psalms* is perhaps the ultimate polysemic text, especially when it comes to its views of nature and the environment. It can be read as either promoting the most egregious and destructive forms of objectification and ownership of nature and the environment, or as supporting the most environmentally friendly and progressive forms of environmental sustainability, especially when it comes to pastoralism and agriculture in which a beneficent God blesses the earth and His chosen people. The latter reading is especially the case with the revisionist readings of some of the *Psalms* undertaken by the Earth Bible Team who have set about the task of 'the re-earthing of the Bible' (Geering, 2001, p. 13) as part of the general movement of 'ecotheology'.

Rather than seeking to recuperate and to conduct a revisionist reading of the *Psalms* for ecology (as the Earth Bible Team does that, selectively discussing aspects of the *Psalms* that support its project and ignoring or overlooking those that don't), the ecocritical reading I undertake in this chapter for conservation and green politics teases out some of the contradictory and problematic views of nature, the earth, humans, plants, animals, land, water and God, and their relationships with each other in the *Psalms* and readings of them, including the Earth Bible Team's reading of some verses of the *Psalms* and C. S. Lewis's reading in *Reflections on the Psalms.* In this chapter I argue that the Psalmists' views of nature and the environment, and some of their later readers in the Earth Bible Team, support a destructive environmental ecotheology underpinned by Western philosophical and political concepts of subject and object with relationships of ownership and enslaving by the former of the latter. In response and instead, I argue for an environmental or earthing theology of abjects in an inter-corporeal relationship between the sacred body of the earth and the sacred human body. This relationship would entail nurturing mutuality and symbiosis with the earth and its creatures, especially water beings who generally get short shrift in the Old Testament (as we have already seen and in which the *Psalms* participate).

God the dominator

The Earth Bible Team espouses 'Six Ecojustice Principles'. The third of these principles, the 'principle of voice' proclaims that 'Earth is a subject'. In addition, 'Earth is a living entity', 'an all-embracing organism' (Earth Bible Team, 2000, p. 46) and 'a living subject, as distinct from a lifeless object' (Earth Bible Team, 2001, pp. 22 and 24). God is also a subject, as the Earth Bible Team maintains (2001, p. 24), yet for God the earth is an object, including an object of ownership as announced in *Psalm* 24: 1: 'the earth is the Lord's'. Lewis (1958) and Horrell (2010) ignore this verse and the Earth Bible Team pass over it without comment (Trudinger, 2001, p. 37). The earth is also an object of action upon which God acts firstly by creating it. God and the earth are not subjects of the same status, nor are they in a reciprocal relationship as the earth does not own the Lord and the earth does not act on the Lord.

The leader of the Earth Bible Team, Norman Habel, previously outlined six ideologies of God's ownership of the earth and land in the Old Testament. These include a royal ideology in which God as 'a divine monarch' (or dominator) has 'dominion over the whole earth as an empire' and who views and treats 'the land as the source of wealth' (Habel, 1995, pp. 17 and 27). On this view, God is emperor over the whole earth, God has colonized the whole earth and the land is treated as a source of extractable wealth. God also gave dominion to humans over earth as enunciated in *Genesis* 1: 28 (which Habel does not discuss). Surely the whole earth needs to be decolonized, the land liberated from slavery as a resource of commodities and treated with respect as a living being.

The Earth Bible Team's readings of the *Psalms* generally overlook this royal ideology and other equally egregious land ideologies that Habel had previously outlined and that it underpins, such as the theocratic ideology of God as land-owner (Habel, 1995, pp. 36–53) and the earth as God's entitlement or rightful property to grant land to His chosen people, even if it means dispossessing other people (Habel, 1995, p. 35, 75–96).

One exception among members of the Earth Bible Team who addresses God as dominator in the *Psalms* is Keith Carley in his reading and rewriting of *Psalm* 8, 'widely regarded as one of the most inspiring, uplifting poems ever written' (Carley, 2000, p. 111; for his rewriting see pp. 123–124). Carley (2000, p. 111) admits straight up that 'Earth is portrayed in the psalm as an inanimate stage for the display of divine and human power rather than the vital source of life we now know it to be.' This is wishful thinking as 'we', or at least some members of the human species of the industrial capitalist consumerist sort, still portray and act towards the earth as an inanimate stage for the display of divinely ordained human power by way of resource extraction, industrial exploitation and displays of technological mastery.

Carley (2000, p. 115) also concedes that

> in this psalm, Earth is voiceless! There is no engagement with the ecojustice principle of voice. We have no indication of Earth's response to the psalmist's celebration of God's power throughout its length and breadth [. . .] The theocentric and anthropocentric foci of the psalm leave no room for the voices of either the Earth or its non-human creatures.

All texts, however, leave room for the voices of later commentators, and this psalm is no exception. The New Testament writer of the book of *Hebrews* and

C. S. Lewis both comment on this psalm. Their voices, in turn, may leave room for the voice of the earth and all its creatures, human and non-human alike.

Referring to *Psalm* 8, for C. S. Lewis (1958, p. 132) 'this short, exquisite lyric is simplicity itself – an expression of wonder at man and man's place in Nature [. . .] and therefore at God who appointed it'. Lewis (1958, pp. 132–133) points out that the claim made in this *Psalm* that God has 'made us lords of all the other creatures [. . .] is not strictly true' as 'man is often killed, and still more often defeated, by beasts, poisonous vegetables [*sic*], weather, earthquakes'. Nature is not necessarily benign and docile, but is sometimes indomitable and dominating. In support of this view Lewis refers to the writer of the New Testament book of *Hebrews* (2: 6–9) commenting on this *Psalm*. But 'the Christian writer', as Lewis calls him, of *Hebrews* does not elaborate as Lewis does, nor draw the same conclusions as Lewis does who nevertheless accords 'Nature' or 'Earth' and its creatures, including plants (albeit poisonous) and animals (albeit 'beasts'), 'voice', or at least agency.

Unlike Lewis, the Earth Bible Team never uses the old-fashioned term 'Nature', though Habel surreptitiously re-introduces lower case 'nature' more recently in writing about ecological hermeneutics as if he could not keep nature at bay despite his best efforts to do so (Habel, 2008, p. 3). Like dragons, nature seems to keep on raising its beautiful (or otherwise) head. The Earth Bible Team generally prefer 'Earth' to Nature and nature. Habel and the Earth Bible Team translate the Hebrew *Erets* as 'Earth' (Habel, 2000, pp. 35, 36; Earth Bible Team, 2000, p. 27), and not as 'the earth', nor as 'land'. The Hebrew term *Erets* 'functions to refer both to 'earth' and 'land' (Brueggemann (2002, p. xiii) see also Habel (1995, p. 2).

More recently Habel (2008, p. 3) surreptitiously re-introduces 'nature' by using the term 'Earth' to refer to 'the total ecosystem, that is, the web of life – the domain of nature'. The Earth Bible Team tries to avoid objectification by deleting the definite article when referring to 'Earth', but does not avoid it by invoking the domain of nature, which implies ownership of a territory by a ruler. Which ruler being referred to here is not clear in this context – does God, or 'Earth', or humans, or nature, own nature? Whose domain is nature? Are 'Earth' or nature capable of self-ownership? Earth and nature are capable of self-determination to some degree, or agency.

Habel's use of 'domain' also implies dominion (and a dominator) and so alludes to *Genesis* 1: 28 in which God gives humans the authority and power (as subjects) to subdue and have dominion over earth and all its creatures (as objects). As God gave humans this role before the Fall, did humans lose the right to exercise this dominion by sinning? Did humans then become subjects like earth? With which we might engage in an inter-subjective relationship? The Earth Bible Team does not also avoid problematic subjectification in which capitalised 'E' 'Earth' becomes equivalent as a subject to capital 'G' God. I prefer lower case 'e' 'earth' without a definite article to signify both subject and object, or better still neither – earth as abject to use Julia Kristeva's term (discussed below).

Earth is a subject as earth also acts and sings praises to God (as announced in many *Psalms* and as enumerated by the Earth Bible Team (2001, p. 27)). The nature of the relationship between the two subjects of God and earth is generally only considered to be inter-subjective by the Earth Bible Team, yet they are two different classes of subject with different capabilities as God created earth (earth did not create God), as earth is an object of ownership by God (earth does not own God), and as earth sings praises to God (God does not sing praises to earth). Both earth and God may be subjects, or may be considered as subjects, but the relationship between them is not a reciprocal or dialogic relationship between two equivalent subjects. It is a monologic relationship in which one created and owns the other, and in which they address messages to each other, but they don't necessarily hear or listen to each other. In some ways, they talk past each other.

Declaring the glory of God

Psalm 19 is a case in point. Lewis (1958, p. 63) 'take[s] this to be the greatest poem in the Psalter and one of the greatest lyrics in the world'. The Earth Bible Team inexplicably does not consider this *Psalm* at all, yet it is about the voice of the earth and its relationship to God. In *Psalm* 19: 1–4:

> The heavens declare the glory of God; the skies proclaim the work of his hands. Day after day they pour forth speech; night after night they reveal knowledge. They have no speech, they use no words; no sound is heard from them. Yet their voice goes out into all the earth, their words to the ends of the world.

The heavens and the skies paradoxically pour forth speech, but they do not have speech. Perhaps they are a mere conduit for speech without having it or producing it themselves. Speech passes through them. They use no words, but they convey words. They have voice (which supports the Earth Bible Team's third principle of ecojustice; Habel (2011, p. 22)), but they have no sound (which does not support it). They have soundless voice. They communicate, but not in language, for those who have ears to ear. They are abjects without significance who engage in *signifiance* (to use another of Kristeva's terms also discussed below).

Whether God acknowledges, recognizes or receives these actions and hears these communications emitted or produced by earth is another matter, for God is a mystery. God is certainly in communication with earth, but whether earth is in communication with God is another question altogether. Earth certainly addresses and sends messages to God, but whether God receives them no one knows for God is in monologue, not dialogue, with earth. He is (only) in dialogue with humans as beings endowed with language (in biblical terms) and to whom God speaks (and who speak to God). He never replies to earth, such as responding to its/her praises of Him. Earth reflects God as a passive mirror giving Him back His glory, not as an active agent acting upon God.

Humans, as part of God's creation of earth, are also objects of ownership by God, as are all of earth's inhabitants. *Psalm* 24: 1 goes on to announce that 'the earth is the Lord's, all that is in it [. . .] and those who live in it'. Humans are both objects owned by God and subjects endowed with language by God who can and do address God. Humans are caught in this philosophical and theological conundrum of being subjects (like God) endowed with language, and being objects (like earth, but in dialogue with God). Humans are objects for God and subjects that communicate with God and the rest of creation, which is also able to address God, sing His praises, etc., according to many of the *Psalms*.

Grammatically, if there is a subject, there is also a predicate, usually comprising an object. God is the objectless subject: 'I am'. God is also a subject who self-predicates: 'I am that I am' (*Exodus* 3: 14). God is also a subject with a copula and an object: 'God is love'. Subject and predicate are mutually defining and regarding entities. Philosophically, though, the distinction between subject and predicate sets up subjects and predicates. The distinction between them precedes them and makes them possible. Friedrich Nietzsche suggested that philosophy before him began with 'the subject', a thinking substance, or 'I' as in Descartes' famous *cogito ergo sum*, 'I think, therefore I am'. Being for Descartes is predicated of the subject who thinks. Without thinking, there is no subject. Theology for the Earth Bible Team begins with the subject. Yet the subject for Nietzsche is a back formation and invention *post hoc* (after the event). In §481 of *The Will to Power* he argues that 'the 'subject' is not something given, it is something added and invented and *projected behind what there is'* (my emphasis). Who, or what adds, invents and projects the subject behind what there is? The answer for Nietzsche is simple: the body. The body is the starting point. Before we are subjects, we are bodies.

The Earth Bible Team projects God, Earth and humans as subjects behind what there is. This projection operates along similar lines to Thomas Aquinas' 'first cause' and 'necessary being' proofs for the existence of God, except that for the Earth Bible Team Earth and humans are also subjects with God. God, however, is a different class of subject to the Earth and humans as He created them. He is neither created, nor self-creating. He is a self-actualizing, self-predicating subject: 'I am' and 'I am that I am'.

Philosophically, though, subjects are not necessarily different in status from other subjects, don't necessarily have to regard and treat others as objects and can interact with other subjects inter-subjectively (which is what the Earth Bible Team wants to achieve between God and earth, but in doing so they arrogate to 'Earth' greater subjectivity than it can bear theologically and linguistically For them capital 'E' Earth becomes equal or equivalent to God, a heresy). Linguistically, if there is a subject, there can either be a second person (human) subject ('I love you') capable of replying ('I love you too'), and/or a third person object ('I love earth') that cannot reply and address human subjects as 'you'.

An inter-subjective relationship between humans and earth would mean mutually reciprocal communication between both of us with humans addressing the earth ('I love you/ earth') without it being clear what you/ earth might say in reply (perhaps, 'you are killing me'). Plants and animals do communicate between members of their own species and other species. Humans listening to members of other species has occurred for generations without humans necessarily being addressed as 'you' by other species.

Philosophically and linguistically, Julia Kristeva's concept of the abject deconstructs and decolonizes the subject-object distinction altogether as it precedes and refuses it (Kristeva, 1982, pp. 1–2). Like the body for Nietzsche, the abject is what there is. An environmental theology of abjects (God, or the Goddess, as (the) abject; humans and earth as abjects), would look quite different from the Earth Bible Team's ecotheology of subjects and inter-subjectivity. There would be no creation of one by the other, no objectification of one or the other, no ownership of one by the other, no subjects and objects. There would be co-evolution of each with the other and the singing of the praises of many by many others. The earth in an environmental theology of abjects is a material and spiritual living being. Earth, land and body are abjects. Traditional cultures figure earth and land as body, and/or figure body as earth and land (see Giblett, 2008, especially chapters 10 and 11). This figuring entails an inter-corporeal relationship of loving union between the sacred body of the earth and the sacred human body (as we will see in the final chapter).

Politically, subjects are subjects of a sovereign, and this is certainly the case with humans in the royal and theocratic land ideologies. In the case of humans and earth as subjects, they are subjects of and to the Lord God. He as the sovereign subject is not subject to anyone or anything else. Legally, subjects have varying statuses and rights depending on the political and legal regime under consideration. Earth has not tended to have the same rights as humans in Western culture. Earth as subject has been an object of ownership and exploitation, and so has not had personhood. Persons are not objects of ownership, except in slave and feudal societies. Earth is also enslaved in royal and theocratic land ideologies. Earth until recently has only ever had personhood in indigenous cultures.

In Western jurisprudence earth has generally not had personhood until very recently. *The Guardian* newspaper reported recently that, 'in a world first a New Zealand river has been give the same legal rights as a human being' based on traditional indigenous Maori concepts of personhood and legal status.[1] To regard earth as subject would mean granting it the same legal rights as human subjects. Yet biblically earth and rivers, for instance, do not have the same status, rights, responsibilities and privileges as they do in indigenous cultures. For instance, the still waters of *Psalm* 23: 2 paint a pretty, pastoral picture of the Lord's beneficence, but they do not have the same status as the first-person Psalmist who the Lord leads by them, nor of the Lord who does the leading. The Psalmist is the active subject and the still waters are a passive object. They do not act, but are acted upon.

Nature in the *Psalms*

Earth and land are problematic in the *Psalms*, to put it bluntly. In the late 1950s C. S. Lewis addressed head on the problem of the view of what he called

'Nature' in the *Psalms*, though he does not define the term 'Nature' and he conveniently ignores or overlooks *Psalms* 23: 2 and 24: 1 (as does Horrell, 2010). I will take nature to mean a collective noun for earth, land, air, fire, water, plants, animals, planetary motion, solar energy, climate, the weather, the seasons, etc. In his *Reflections on the Psalms* in the chapter on 'Nature' Lewis argues that the Jews had two contending views of nature. The first was that they lived 'close to the land' and were 'vividly aware of our dependence on soil and weather' (Lewis, 1958, p. 76). Those who live close to the land tend to have little regard or time, then and now, for landscape, land considered as something to be viewed alone, and not worked (as Raymond Williams (1973, p. 120) said in *The Country and the City*). The Psalmists are no exception as they for Lewis (1958, p. 77) 'naturally give us little landscape', though when they do so, such as in *Psalm* 23 with its green pastures, still waters, and the valley of the shadow of death, these landscapes are pressed into allegorical service to make a theological point and are not being appreciated for their own sake, nor aesthetically for the pleasure they may give.

Those who live close to the land do not usually look at the land as landscape, but they invariably have an eye on the weather as that affects when and how they work the land, when they can sow and reap, what the land does and how much it yields. Lewis remarks that what the Psalmists give us, 'far more sensuously and delightedly than anything I have seen in Greek, is the very feel of the weather' (Lewis, 1958, p. 77). Lewis goes on to cite in support *Psalm* 65: 9–13 addressed to God:

> You care for the land and water it; you enrich it abundantly. The streams of God are filled with water to provide the people with grain, for so you have ordained it. You drench its furrows and level its ridges; you soften it with showers and bless its crops. You crown the year with your bounty, and your carts overflow with abundance. The grasslands of the wilderness overflow; the hills are clothed with gladness. The meadows are covered with flocks and the valleys are mantled with grain; they shout for joy and sing.[2]

The God of these verses is an agricultural and pastoral God whose beneficent meteorological activity is praised in terms of the benefits He brings to the land, crops and pastures. These benefits in turn praise God. The land referred to in this *Psalm* is the promised land flowing with milk and honey, and not the blighted land drowning in floods, nor the parched land baked in drought, nor the fertile wet lands of marshes and swamps. This is also the land of gentle rain, not violent storms. Pastoral ideology is a largely distinct biblical land ideology that has obvious pertinence to European culture and its settler diasporas (see Giblett, 2011, chapters 4 and 10).

In pastoral ideology 'the land is good' that Habel (1995, p. 41) briefly touches upon referencing and discussing *Deuteronomy*, but the good land is a gift from God (Habel, 1995, p. 43; Brueggemann, 2002, p. 45), and is not (a) good in, of or for itself. 'The land is for satiation', as Brueggemann (2002 46) puts it, for and by God's chosen people, and not for all people. Nor is the land for conservation to ensure its ongoing ability and capacity to satisfy the needs of all people, God's chosen ones and those not chosen. The good land contrasts with the bad lands of the wilderness, of 'harsh and alien desert lands', as Habel (1995, p. 78) puts it, or 'landlessness [. . .] a place without resources', as Brueggemann (2002, pp. xi, 40) puts it. Wilderness ideology is another largely distinct biblical land ideology that has obvious pertinence to the United States and Australia and their wilderness areas and associated legislation (see Giblett, 2011, chapter 5).

According to Lewis (1958, p. 77), the second view of nature for the Jews is that 'Nature and God were distinct; the One had made the other; the One ruled and the other obeyed'. Nature on this view is not only *not* divine, but it is also a made object, not a making subject; it is a subject in the political sense of being ruled over and ordained to obey. Referring to Lewis, Walker-Jones acknowledges that 'the separation of God from nature intrinsic to much Western theology has facilitated the colonization of women, indigenous people and earth' (Walker-Jones, 2001, p. 87). The Earth Bible Team tries to rethink the relationship between God and earth ostensibly from within Judeo-Christian theology, but does so in fact from within Western philosophy with the concept of the subject (alone) in which God, humans and earth are all subjects (and nothing or no one is object or abject) who relate to each other inter-subjectively. Philosophically and linguistically this is problematic, if not impossible. Such a view is aimed towards the decolonization of women, indigenous people and earth/land (as some of the team members attempt in their contribution to this and other volumes in the Earth Bible Team series).

Decolonization of all three would, however, need to address the ways in which earth, women and indigenous people under colonization were not, and under neo-colonization are not, granted the same status and rights as straight white men were and are, and how the former are regarded as less than human and sometimes constituted and treated as objects of ownership by the latter. Decolonization of all three would need to address the ways in which some places of or on earth are regarded as more valuable and useful than others. Less valuable and useful places are also places where indigenous peoples live on the outskirts, or in the inner slums, or in the lower regions of the colonial centres of power and wealth in the upper regions. Colonization of women, indigenous people and earth are tied up with each other, often one being used to figure the other, such as regarding swamps as female and vice versa. Decolonization disentangles them.

Lewis (1958, p. 80) goes on to point out that 'to say that God created Nature, while it brings God and Nature into relation, [it] also separates them. What makes and what is made must be two, not one. Thus the doctrine of Creation in one sense empties Nature of divinity'. This is the only occasion on which Lewis is cited in the Earth Bible Team's book on the *Psalms* (see Walker-Jones, 2001, p. 87). This seems strange as Lewis comments extensively on nature in the *Psalms* and devoted a whole chapter to 'Nature' (as we have seen and will continue to see). Both the counter-doctrine of the Goddess and nature being mutually self-creating, and the cognate counter-doctrine of co-evolution, fill nature with divinity as sacred earth. Mutuality and co-evolution entail an inter-corporeal relationship between the sacred body of the earth and the sacred human body. What makes and what is made are one, not two. A procreating couple conceiving a child is one, not one, or two, or three. The one is making a made within the union of the one.

In the second sense of the doctrine of creation for Lewis (1958, p. 81), 'the same doctrine which empties Nature of her divinity also makes her an index, a symbol, a manifestation of the Divine'. Walker-Jones (2001) does not cite Lewis's discussion of a feminized 'Nature' (and so does not address the concept of 'Nature' as an index, a symbol, a manifestation of 'the Divine'). Lewis feminizes 'Nature', but there is no biblical warrant for doing so, certainly not in the *Psalms*. Perhaps, then, it is not surprising that Walker-Jones does not cite Lewis here, but by doing so he ignores or overlooks that 'Nature' is and has been feminized and divinized (as Goddess). God is masculine or masculinized, and there is plenty of biblical warrant for doing so, so perhaps 'Nature' is thereby implicitly feminized and Lewis seems to readily accede without comment to that implication. To say that 'Nature' indicates, symbolizes or manifests God, while it brings God and Nature into relation, also separates and mediates them. Masculinizing God and feminizing 'Nature' also places them on either side of a gendered divide irrespective of the relationship that may ensue between them, though it has been normatively patriarchal and heterosexual, not otherwise and not transgendered. Other relations – sexual and political – are not embraced and included in the earthly polity both in the *Psalms* and elsewhere in the Bible.

For Lewis, this mediation means that 'because natural objects are no longer taken to be themselves Divine that they can now be magnificent symbols of Divinity' (Lewis, 1958, p. 81). Natural objects are not emptied totally of divinity, but only have divinity insofar as they indicate, symbolize or manifest God. If they don't perform these actions or fulfil these roles, then they don't have divinity. Natural objects are instrumentalized in the service of God and made to signify for the semiosis of God. They have no divine significance or value otherwise.

In Kristeva's terms, natural objects for Lewis have significance, but they do not have or perform *signifiance*, playful, nonsensical poesis (or making) (Kristeva, 1984, p. 17). Green pastures and still waters perform positive and significant ecosystem and semiotic services and aesthetic (from *aesthesis*, meaning sensing, feeling) functions, but 'the valley of the shadow of death' (*Psalm* 23: 4) does not, or it does so only with a negative capability. It is not really a natural object as it is not a natural place or process, unlike green pastures and still waters. A valley is a natural place and death is a natural process, but the valley of the shadow of death is neither. It manifests and symbolizes death and evil, and by doing so adumbrates life in contrast to death and good in contrast to evil, and God as the Being of both life and good.

Lewis (1958, p. 81) concludes that 'the doctrine of Creation leaves Nature full of manifestations which show the presence of God, and created energies which serve Him'. This doctrine does not, however, leave Nature full of divinity and energies that serve the earth and all its inhabitants. In fact, it empties Nature of divinity (as Lewis indicated previously). Nature on Lewis's view is a servile servant, subject politically and ecologically to God who is 'the source of Earth's life and abundance', as Wallace (2001, p. 57) puts it when commenting (like Lewis) on the view of God and Earth in *Psalm* 65. Nature is a subject, if not a slave, in the service of God first, and humans second (as famously enunciated in *Genesis* 1: 28). Feminized nature on this view lies fallow and fertile until masculinized beings (human and otherwise) inseminate it. Then it brings forth life. Aristotle had a similar patriarchal view of nature, procreation and women's bodies (see Giblett, 1996, pp. 156–160).

Rider on the storm

Rather than *Psalm* 65, *Psalm* 104 for Lewis (1958, pp. 83–84) is 'the great *Psalm* especially devoted to Nature'. Walker-Jones concurs by concluding his reading of this *Psalm* that 'Earth is neither background nor object but an organic, living, central subject of *Psalm* 104' (Walker-Jones, 2001, p. 97; see also Ntreh, 2001, pp. 98–108; Horrell, 2010, pp. 49–61). Yet rather than earth, the Lord is arguably the central subject in this *Psalm* as the very first two verses announce, though He is not organic (Walker-Jones does not clearly differentiate, nor enunciate the relationship, between the subjects of earth and the Lord). The Lord of *Psalm* 104: 1 and 2 is 'very great, clothed with splendor and majesty and who wraps himself in light as with a garment and stretches out the heavens like a tent'. The Lord is light in a kind of super-constellation in the form of an overarching nomadic pastoral tent erected over the earth that contains and protects it. He endorses the nomadic pastoral way of life in a simile as a tent writ large in a constellation. This emperor's clothes are plain for all to see and His nakedness is not seen at all.

The Lord is not only astronomical, but also meteorological. In *Psalm* 104: 3 and 4 He 'makes the clouds his chariot and rides on the wings of the wind. He makes winds his messengers, flames of fire his servants'. Similarly, in *Psalm* 68: 4 He 'rides on the clouds'. Walker-Jones (2001, p. 89) comments on *Psalm* 104 that the Lord is 'a storm god'. The Lord is a storm god and sky lord, not an earth, nor land, nor water Lord. He is lord over and of earth, land and water as He is outside and above them. Similarly, in *Psalm* 29: 10 He 'sits enthroned over the flood'. He transcends and reigns over water and its beings. He is also a pyrotechnical, meteorological and military lord (or 'a warrior god', as Habel (1995, p. 73) calls Him) as He sends wind-borne messages, commands fire and rides in a chariot. He also sends messages on the wind. This depiction both harks back to the misreading of the Spirit of God in *Genesis* 1: 2 hovering over land and water before they were separated and points forward to the Holy Spirit of the *Acts of the Apostles* 2: 2–3 riding on the wind and speaking in tongues of fire.

The Lord of *Psalm* 104: 3 is a builder too as He 'lays the beams of his upper chambers on their waters'. Similarly, the Lord of *Psalm* 24: 2 'founded it [the earth] on the seas and established it on the waters'. The earth has a marine, or aqueous foundation, or even an aquiferous foundation (founded on aquifers) and is aqua-terrous (wet land) (*Genesis* 2: 6). The Lord of *Psalm* 104: 6 'covered it [earth] with the watery depths as with a garment'. The Lord is clothed in stars while earth is clothed in water. The two subjects are quite distinct in this regard, though they share the common characteristic of being clothed. No nakedness here please! The Lord is the subject who acts, and who acts upon earth (and not vice versa). Earth certainly acts, but not upon the Lord. Earth is that which is acted upon by the Lord. They are not equivalent subjects who could have a mutually reciprocal relationship.

The Lord controls land and water as a kind of super-civil and -hydraulic engineer who acts upon earth, including land and water (earth does not engineer the Lord). In *Psalm* 104: 5–10:

> He set the earth on its foundations; it can never be moved [. . .] the waters stood above the mountains. But at your rebuke the waters fled, at the sound of your thunder they took to flight; they flowed over the mountains, they went down into the valleys, to the place you assigned for them. You set a boundary they cannot cross; never again will they cover the earth. He makes springs pour water into the ravines; it flows between the mountains.

The Lord divided the water from the water (*Genesis* 1: 6). Some theologians, such as John Day and David Horrell, have drawn a parallel between *Psalm* 104: 5–9 and *Genesis* 1: 6–10 as in both passages 'waters [are] pushed back' (Horrell, 2010, p. 50, citing John Day). Wetlands on this view as undivided waters and lands are an aberration or a throwback to *Genesis* 1: 2 when 'the earth was formless and empty', when water crossed the boundary with land. Civil and hydraulic engineering to manage water is another distinct biblical land ideology that has obvious pertinence to Europeans and their settler diaspora's view and drainage of wetlands, but not one identified by Habel (1995; see also Giblett, 1996; 2016).

Horrell (2010, p. 57) also draws a parallel between *Psalm* 104: 6, 9 and *Job* 38: 8 in which the Lord 'shut in the sea with doors when it burst out from the womb'. This verse acknowledges the fertile and creative qualities of water while emphasizing that the Lord controls the flow of water. He does so to and for the benefit of His chosen people and their domesticated animals alike. The mountains in *Psalm* 104: 11–15:

> give water to all the beasts of the field; the wild donkeys quench their thirst. The birds of the sky nest by the waters; they sing among the branches. He waters the mountains from his upper chambers; the land is satisfied by the fruit of his work. He makes grass grow for the cattle, and plants for people to cultivate – bringing forth food from the earth: wine that gladdens human hearts, oil to make their faces shine, and bread that sustains their hearts.

Land lacks water so the Lord supplies water and satisfies thirst. The Lord also supplies grass and grains for domesticated animals and for the pastoral, horticultural, viti-cultural, oenological, bread-sustained, wine-drinking and heart-gladdened people.

The Lord also supplies water to non-domesticated plants, such as trees, and to non-domesticated animals, such as birds. In *Psalm* 104: 16–18:

> The trees of the Lord are well watered, the cedars of Lebanon that he planted. There the birds make their nests; the stork has its home in the junipers. The high mountains belong to the wild goats; the crags are a refuge for the hyrax.

The Lord also fits habitat to species and vice versa as an engineer of co-creation, to make a home for all and to make every place a home for some creature, though the creatures here are land and sky creatures, not water creatures (which can be problematic as we will see shortly).

As a meteorological and astronomical being in *Psalm* 104: 19–23:

> He made the moon to mark the seasons, and the sun knows when to go down. You bring darkness, it becomes night, and all the beasts of the forest prowl. The lions roar for their prey and seek their food from God. The sun rises, and they steal away; they return and lie down in their dens. Then people go out to their work, to their labour until evening.

The revolution of the moon around the earth and of the earth around the sun (and their determination of the cycles of day and night, the tides and the seasons) are outside the knowledge of the Psalmist and are ascribed to the Lord as some of His many splendid works. Meteorological and astronomical phenomenon are neither random nor chaotic, but are ordered and controlled by the Lord as part of the biblical pastoral land ideology.

Water beings in the *Psalms*

The Lord is a worker like the people. In *Psalm* 104: 24–26 the Psalmist proclaims:

> How many are your works, Lord! In wisdom you made them all; the earth is full of your creatures. There is the sea, vast and spacious, teeming with creatures beyond number— living things both large and small. There the ships go to and fro, and Leviathan, which you formed to frolic there.

While Leviathan is here a sea-creature and probably a whale, for other biblical writers Leviathan is not such a benign and frolicsome creature (as Walker-Jones (2001, pp. 90–92) and Beal (2002, pp. 26–27) point out and as we have seen in previous chapters). Leviathan in *Psalm* 74: 13 is associated by parallelism ('the practice of saying the same thing twice in different words' as C. S. Lewis defines it (1958, p. 3)) with 'the dragons of the waters', an apt description of a crocodile (as we saw in the previous chapter). In other words, and for some biblical writers, Leviathan is a crocodile (as we also saw in the previous chapter). The crocodile might be regarded as a dragon of the waters as the crocodile breathes out hot and wet breath through its nostrils.

Trudinger comments on *Psalm* 74 that:

> In the psalm, great importance is given to Earth. In the guise of the temple, it is regarded as treasured by God; as cosmos, it is a place of harmony and order. Divine concern for Earth, it is assumed, will most certainly result in intervention to control hostile forces in the present. The water beings do not enjoy such a happy status. They are portrayed as God's opponents, suffer horrible injury and a degrading end. They are compared with people who treat God's beloved temple-Earth as if it were only a forest for clearing. In *Psalm* 74, the water-beings are beyond redemption.
>
> (Trundinger, 2001, p. 35)

Water beings are part or aspects of fallen creation created after the Fall when evil entered the world. They are monsters created by Cain, at least if *Beowulf* is anything to go by (as we saw in a previous chapter). They are not part of God's creation and were not created by God. God did not create all living beings and creatures. Water beings embody a patriarchal horror of the watery, the slimy and the monstrous. People who treat the sacred earth as if it were only a forest for clearing are horrible full stop. They should not be acknowledged by comparing them with anything or any other being, certainly not with water beings who live on the earth with land beings.

Walker-Jones (2001, p. 91 and n.10) comments that in *Psalm* 104: 26 'the serpent Leviathan is created to play in the ocean' and selectively cites in a footnote *Isaiah* 27: 1 in which Leviathan is a serpent. Yet this verse more explicitly invokes 'Leviathan the gliding serpent, Leviathan the coiling serpent; he [the Lord] will slay the monster of the sea'. Walker-Jones (2001, p. 94) later comments on *Psalm* 104 that 'Leviathan is not a primeval monster, but was created by God to laugh and play in the sea'. Why, then, will He slay Leviathan as Isaiah prophesizes? Why will He kill one of His own creatures? Walker-Jones contradicts Isaiah and Trudinger for whom Leviathan is God's opponent. Leviathan as a whale may have been created by God to laugh and play in the sea, but Leviathan as a dragon of the waters or monstrous serpent was not created by God to laugh and play in the sea, may not have been created by God at all, may have been created by Cain to oppose and fight God, may not laugh and play, but may kill and destroy (if *Beowulf* is anything to go by), and may not do so in the sea, but in monstrous marshes (also if *Beowulf* is anything to go by), though God will subdue Leviathan (if *Isaiah* is anything to go by). *Isaiah* 27: 1 prophesizes that 'the Lord will punish [Leviathan] with his sword – his fierce, great and powerful sword'. The Lord is a precursor of Beowulf and Saint George, a sword-wielding monster-slayer on land and in water. The Lord is the prototypical monster-slayer. He gives legitimacy to Beowulf and Saint George. They are only following in His footsteps as His obedient servants and exercisers of His will.

Life and death in the *Psalms*

The Lord is also the giver of food, and the giver and taker of life. The Lord is generous, but also terrifying. *Psalm* 104: 27–30 relates how:

> All creatures look to you to give them their food at the proper time. When you give it to them, they gather it up; when you open your hand, they are satisfied with good things. When you hide your face, they are terrified; when you take away their breath, they die and return to the dust. When you send your Spirit, they are created, and you renew the face of the ground.

Hiebert (2008, p. 18) sees this translation of the same biblical Hebrew רוּחַ in verse 29 as 'breath' and in verse 30 as 'Spirit' as illustrating 'post-biblical theological dualism' between spirituality and materiality (as we saw in the first chapter).

In this biblical verse, the ground is humanized as having a face and a surface. Whether earth has depths, whether these are renewed by the Lord and whether earth is a body are other questions. Walker-Jones (2001, p. 91) comments that in *Psalm* 104: 29–30 'death is not viewed as something evil, but as a natural part of the cycle of life'. Yet in these verses human death is figured as humans returning to the dust from which they/we are made (*Genesis* 2: 7) in accordance with the Lord's curse on humans of *Genesis* 3: 19 for disobeying His command not to eat of the tree of the knowledge of good and evil: 'dust you are and to dust you will return.' Death is regarded as the end. No new life springs from this dust. The cycle of life here is dust-life-dust and not the circle of life-death-new life in which new life springs out of death. Dust is a tomb, unlike the watery womb, or womby waters.

Psalm 104: 31–32 concludes: 'May the glory of the Lord endure forever; may the Lord rejoice in his works – he who looks at the earth, and it trembles, who touches the mountains, and they smoke.' The Lord is also a being who induces the seismic phenomenon of earth through the power of his gaze and touch. Volcanic eruptions and earthquakes are His doing. The Lord is not a being of the wetlands where earth trembles; the Lord is not a being of the native quaking zones of the swamps.

The Psalmists' views of nature (or earth, or land) are problematic to say the least, and the various readings of the *Psalms* encountered and explored above highlight some of the problems with these views. They highlight especially the culturally and historically contingent nature of those views (not that mine are any less so, though there seems more reluctance on the part of other readers to acknowledge this possibility). The problematic nature (in two senses) of the Psalmists' environmental theology creates a place to nurture an environmentally friendly or earth theology of earth/Goddess. This would sense, respect and act towards nature as a being of sky, storms, waters and earth, of animals and plants domesticated and wild, of sky, land and water beings, with whom and about whom to celebrate life in dance, song, word and tai chi in an inter-corporeal relationship of loving union between the sacred body of the earth and the sacred human body.

Notes

1 Eleanor Ainge Roy, 'New Zealand river granted same legal rights as human being' [*sic*], *The Guardian Weekly*, 24.03.17, p. 12. Also available online at: https://www.theguardian.com/world/2017/mar/16/new-zealand-river-granted-same-legal-rights-as-human-being
Two days later a court in India, citing the New Zealand case and verdict as a precedent, did the same with two Indian rivers. See Michael Safi and agencies, 'Ganges and

Yamuna rivers granted same legal rights as human beings'. Available online at: https://www.theguardian.com/world/2017/mar/21/ganges-and-yamuna-rivers-granted-same-legal-rights-as-human-beings

2 For a member of the Earth Bible Team's ecotheological reading of this *Psalm* see Wallace (2001, pp. 51–64).

References

Beal, T. 2002. *Religion and its Monsters*, New York: Routledge.

Brueggemann, W. 2002. *The Land: Place as Gift, Promise, and Challenge in Biblical Faith*. 2nd Edition. Minneapolis: Fortress Press.

Carley, K. 2000. *Psalm* 8: An Apology for Domination. *In*: N. Habel, ed., *Readings from the Perspective of Earth*, Sheffield: Sheffield Academic Press, pp. 111–124.

Earth Bible Team, The. 2000. Guiding Ecojustice Principles. *In*: N. Habel, ed., *Readings from the Perspective of Earth*, Sheffield: Sheffield Academic Press, pp. 38–53.

Earth Bible Team, The. 2001. The Voice of Earth: More than Metaphor. *In*: N. Habel, ed., *The Earth Story in the Psalms and the Prophets*, Sheffield: Sheffield Academic Press, pp. 23–28.

Geering, L. 2001. Preface. *In*: N. Habel, ed., *The Earth Story in the Psalms and the Prophets*, Sheffield: Sheffield Academic Press, pp. 13–15.

Giblett, R. 1996. *Postmodern Wetlands: Culture, History, Ecology*. Edinburgh: Edinburgh University Press.

Giblett, R. 2008. *The Body of Nature and Culture*. Basingstoke: Palgrave Macmillan.

Giblett, R. 2011. *People and Places of Nature and Culture*. Bristol: Intellect Books.

Giblett, R. 2016. *Cities and Wetlands: The Return of the Repressed in Nature and Culture*. London: Bloomsbury.

Habel, N. 1995. *The Land is Mine: Six Biblical Land Ideologies*. Minneapolis: Fortress Press.

Habel, N. 2000. Geophany: The Earth Story in *Genesis* 1. *In*: N. Habel and S. Wurst, eds, *The Earth Story in Genesis*, Sheffield: Sheffield Academic Press, pp. 34–48.

Habel, N. 2008. Introducing Ecological Hermeneutics. *In*: N. Habel and P. Trudinger, eds, *Exploring Ecological Hermeneutics*, Atlanta: Society of Biblical Literature, pp. 1–8.

Habel, N. 2011. *The Birth, the Curse and the Greening of the Earth: An Ecological Reading of Genesis 1-11*. Sheffield: Sheffield Phoenix Press.

Hiebert, T. 2008. Air, the First Sacred Thing: The Conception of רוּחַ in the Hebrew Scriptures. *In*: N. Habel and P. Trudinger, eds, *Exploring Ecological Hermeneutics*, Atlanta: Society of Biblical Literature, pp. 9–19.

Horrell, D. 2010. *The Bible and the Environment: Towards a Critical Ecological Biblical Theology*. Durham: Acumen.

Kristeva, J. 1982. *Powers of Horror: An Essay on Abjection*. L. Roudiez, trans. New York: Columbia University Press.

Kristeva, J. 1984. *Revolution in Poetic Language*. M. Waller, trans. New York: Columbia University Press.

Lewis, C. S. 1958. *Reflections on the Psalms*. New York: Harcourt.

Ntreh, A. 2001. The Survival of the Earth: An African Reading of *Psalm* 104. *In*: N. Habel, ed., *The Earth Story in the Psalms and the Prophets*, Sheffield: Sheffield Academic Press, pp. 98–108.

Trudinger, P. 2001. Friend or Foe? Earth, Sea and *Chaoskampf* in the *Psalms*. *In*: N. Habel, ed., *The Earth Story in the Psalms and the Prophets*, Sheffield: Sheffield Academic Press, pp. 29–41.

Walker-Jones, A. 2001. *Psalm* 104: A Celebration of the *Vanua*. *In*: N. Habel, ed., *The Earth Story in the Psalms and the Prophets*, Sheffield: Sheffield Academic Press, pp. 84–97.

Wallace, H. 2001. *Jubilate Deo omnis terra*: God and Earth in *Psalm 65*. *In*: N. Habel, ed., *The Earth Story in the Psalms and the Prophets*, Sheffield: Sheffield Academic Press, pp. 51–64.

Williams, R. 1973. *The Country and the City*. London: Chatto and Windus.

Part II

Theologies of times and places

6 Pilgrim's Progress through the Slough of Despond and the Valley of the Shadow of Death

John Bunyan's *Pilgrim's Progress* charts the journey of the pilgrim Christian from the City of Destruction through (among other places) the 'Slough of Despond' and 'the Valley of the Shadow of Death' to the Celestial City. The valley of the shadow of death has a biblical precedent in *Psalm* 23, while the Slough of Despond has no biblical basis or warrant. The association between sloughs and despondency, and wetlands and melancholy more generally, has its roots in the theory of the elements and the humours and its flowers in Robert Burton's *The Anatomy of Melancholy*, first published in 1638, forty years before John Bunyan's *Pilgrim's Progress*. In this chapter I trace the genealogy of the Slough of Despond back to the theory of the elements and the humours (or temperaments) and forward into the nineteenth and twentieth centuries.

The longevity of this genealogy has assisted with the Slough of Despond becoming entrenched as an archetypal landscape in the English psyche, such as 'the grim domains of Giant Despair' in Charles Dickens' *Martin Chuzzlewit*, Blackwater Lake in Wilkie Collins' *The Woman in White*, Great Grimpen Mire in the Sherlock Holmes best-selling detective novel, *The Hound of the Baskervilles*, and in its cultural diasporas, such as late nineteenth-century Melbourne in Fergus Hume's even better-selling detective novel, *The Mystery of a Hansom Cab*. The pejorative figuring of the Slough of Despond also persists into the twentieth-century in popular Christian theological writing, such as C. S. Lewis's *The Pilgrims Regress*, and in American cultural studies, such as Donna Haraway's essay, 'The Promises of Monsters'. I argue that all these writers, like Tolkien, are placist in that they not only ascribe moral qualities to a place undeserving of such opprobrium, but also give them derogatory qualities in the lower psychopathological register. Placism often also ascribes characteristics to a (human-made) place in the city, such as the slums, that were previously ascribed to a place not made with human hands, or made with ancient human hands (jungle, abyss, nether-land, swamps, etc.), such as darkness, decomposition and disease.

Melancholy and madness in the marsh

European culture was founded and still functions on the philosophy of the four elements: earth, air, fire and water (as we saw in the first chapter). Wetlands have always been problematic, indeed aberrant, from this point of view because they mix the elements of earth and water (and even air and 'fire' or heat in the tropics). They also cross the boundaries between land and water, and can even be in transition spatially and temporally between open water to dry land. They are a troubling and unsettling category mediating between land and water.

As wetlands mix the elements, they produce an aberrant 'humour', or psychosomatic state, strictly a kind of phlegmatic melancholy. In the patriarchal Western tradition wetlands have been seen as a wilderness to be tamed, the sites of mixed elements and aberrant humours giving rise to melancholy and madness. The association between melancholy (and depression, despair, despondency, dread and the dismal) and sloughs and other wetlands lies not in the sacred scriptures but in the secular philosophy of the elements and the humours.

In the Western theory of the elements, the four elements combine to form or are associated with psychosomatic states or 'humours' as developed by the Elizabethans and as formulated by E. M. W. Tillyard (1963, p. 76). These states can be tabulated (see Figure 6.1) in terms of the mixing of the four elementary qualities into the four legitimate elements, in turn into four more modern qualitative elements and finally into four proper substances and humours.

This schema becomes problematic when considering the association between melancholia and wetlands, for in the terms of this schema they should not strictly be associated. Or more precisely, wetlands as between or combining earth and water should be seen as equally phlegmatic as they should be seen as melancholic. As a result this combination of elements would produce a phlegmatic melancholy, an aberrant humour.

One of the 'causes of melancholy' for Robert Burton was 'standing waters'. He went on to elaborate in Hippocratic terms that:

> standing waters, thick and ill-coloured, such as come forth of pools and moat where hemp is steeped or slimy fishes live, are most unwholesome, putrefied, and full of mites, creepers, slimy, muddy, unclean, corrupt, impure, by reason of the sun's heat and still standing; they cause foul distemperatures in the body and mind of man, are unfit to make drink of, to dress meat with, or to be used inwardly or outwardly [. . .] So that they that use filthy, standing,

	HOT	COLD
DRY	FIRE HEAT CHOLER/IC	EARTH SOLID MELANCHOLY/IC
'WET'	AIR GASEOUS BLOOD/SANGUINOUS	WATER LIQUID/FLUID PHLEGM/ATIC

Figure 6.1 The Primary Qualities, the Elements and the Humours

> ill-coloured, thick, muddy water, must needs have muddy, ill-coloured, impure and infirm bodies. And because the body works upon the mind, they shall have grosser understandings, dull, foggy, melancholy spirits, and be readily subject to all manner of infirmities.
>
> (Burton, 1932, pp. 224–225).

Although Burton enunciates a psychosomatic theory of the body and mind (after all, he is pre-Descartes and Cartesian dualism), he subscribes to a simplistic analogy, or even correspondence theory, based primarily on colour, between the natural environment and the human body. And although he does not subscribe to the miasmatic theory of physical illness, it is the standing waters themselves, rather than the vapours which rise from them, which are to blame for illness, both 'mental' and physical.

Not only can standing waters cause melancholy, but so also can the cold and dry air which rises from them. When Burton discusses 'bad air' the worst air of all out of hot air, cold air, and cold and dry air is the last which he describes as:

> a thick, cloudy, misty, foggy air, or such as come from fens, moorish grounds, lakes, muckhills, draughts, sinks, where any carcasses or carrion lies, or from whence any stinking fulsome smell comes: [. . .] such air is unwholesome, and engenders melancholy, plagues and what not.
>
> (Burton, 1932, p. 239)

It is on such an 'anatomy', or miasmatic theory, of melancholy and malaria that the Sanitary Movement of two centuries later was founded.

Yet such air is surely not dry but wet. Under the traditional Western theory of the elements and humours as formulated systematically by the Elizabethans, the wetland did not fit easily into the schema (as we have already seen in a previous chapter). In fact, melancholy was strictly associated with the dry upland of the agricultural country and the modern city, and even with capitalism (as Eagleton (1986, p. 41) argues: 'melancholia is an appropriate neurosis for a profit-based society') and with the sublime (as Kant (1960, p. 47; see also 1952, p. 121) postulates: 'the feeling of the sublime is 'sometimes accompanied with a certain dread or melancholy').

In response to the modern melancholic city sublime, there was a massive attempt to displace blame onto the wetland. The Elizabethan 'world picture' as enunciated by E. M. W. Tillyard was not flattering to the modern melancholic and sublimated city-dwelling drylanders; it was more flattering to the pre-modern phlegmatic and 'slimated' fen-dwelling wetlanders. As wetlands did not 'fit the picture', rather than changing the picture, the wetland had to be changed, or more precisely in Burton's case the dwellers in wetlands, including in cities built on or near wetlands (invariably 'foreign'), denigrated.

After enumerating a list of places where the air is bad, Burton (1932, p. 240) concludes by asking 'how can they be excused that have a delicious seat, a pleasant air, and all that nature can afford, and yet through their own nastiness and sluttishness, immund and sordid manner of life, suffer their air to putrefy, and themselves to be choked up?' The wetland, its standing waters, its bad (but wet) airs, and its filthy dwellers, either indigenous wetlanders or city-in-wetland dwellers, is the cause of melancholy displaced from the cold and dry uplands.

The Slough of Despond

Louis Marin described John Bunyan's *Pilgrim's Progress* as 'the most influential religious book ever composed in the English language' (cited by Fritzell 1978, p. 528). It is also the second-most published book in the English language after the Bible. No doubt part of its pre-eminent influence has been to educate generations of readers that an ecologically functioning wetland is not what it seems, and is, but rather an allegorical emblem for a sump of iniquity and cesspool of evil that would drag the unsuspecting and unwary Christian down and entrap them for all eternity. Bunyan (2008, p. 18) relates how 'this miry slough is such a place as cannot be mended; it is the descent whither the scum and filth that attends conviction for sin doth continually run, and therefore it is called the Slough of Despond'.

Things can, and do, go from bad to worse for this pilgrim on his pilgrimage. Wordly-Wiseman advises Christian:

> that Slough is but the beginning of your sorrows, even as other pilgrims experience along the same way [. . .] As you proceed along the way ahead, you are likely to experience wearisomeness, painfulness, hunger, perils, nakedness, sword, lions, dragons, darkness, and in a word, death, and what not?.
>
> (Bunyan, 2008, p. 20–22)

Dragons, darkness and death are also associated in alliterative conjunction. The Valley of the Shadow of Death is dark and a dragon's lair.

Christian encounters a dragon-like monster when he:

> had not gone far in this Valley of Humiliation before he was severely tested, for he noticed a very foul fiend coming over the field to meet him; his name was Apollyon [Destroyer] [. . .] this monster was exceedingly hideous to behold; he was clothed with scales like a fish; (and they are his pride) he had wings like a dragon, feet like a bear, and out of his belly belched forth fire and smoke through a mouth like that of a lion.
>
> (Bunyan, 2008, p. 61)

Apollyon has many typical, dragon-like hybridized features of a swarming creature.

Christian is also a typical dragon-fighter as he has a sword and is a sword-wielder, though he is not a dragon-slayer, unlike both Beowulf and Saint George. In the ensuing battle with the dragon Christian:

> gave Apollyon a deadly thrust [with his sword] which caused him to draw back as if he had received a fatal wound [. . .] As a result, Apollyon quickly spread out his dragon's wings and fled away so that Christian saw him no more. Now unless any man had seen and heard the intensity of this combat as I did, he could not possibly imagine the yelling and hideous roaring of Apollyon, as well as his dragon-like manner of speaking.
>
> (Bunyan, 2008, p. 64)

Apollyon is previously described as having 'a contemptuous, sneering expression', so presumably he spoke in this manner too.

The grim domains of Giant Despair

Despite the increasing secularization of English culture, English writers after Bunyan took up the Slough of Despond as an allegorical emblem of evil and melancholy. The Slough of Despond was divested of its religious overtones, but not of its pejorative, misaquaterrist (wetland-hating) associations. Instead of being a place of evil construed in religious terms, it became a place of melancholia, a kind of secular despondency or despair. Dickens, for example, has 'the Slough of Despond' in *Hard Times*, as well as 'the slough of inanity'. 'The social swamp', as Thomas Huxley (1989, p. 335) called *'la misère'* of the urban underworld of the slums of late nineteenth-century London, could also be 'a slough of despond'.

Dickens, however, saves his most scathing attack on the swamp as a slough of despond for Chapter 23 of *Martin Chuzzlewit*. Dickens subscribed in this novel to a miasmatic theory of melancholia that reproduces the Aristotelian and Elizabethan theory of the elements and the humours. The final leg of the journey to Eden at the confluence of the Mississippi and Ohio Rivers is through a scene made up entirely of:

> sky, wood and water, all the livelong day; and heat that blistered everything it touched. On they toiled through the great solitudes, where the trees upon the banks grew thick and close; and floated in the stream; and held up shrivelled arms from out [*sic*] the river's depths; and slide down the margin of the land, half growing, half decaying in the miry water. On through the weary day and melancholy night: beneath the burning sun, and in the mist and vapour of the evening: on, until return appeared impossible, and restoration to their home a miserable dream.

This is a journey into a place where the elements are mixed, a place which gives birth to death or to a land where the trees are half-growing and half-dying. This mixture of the elements and life and death gives birth to 'the dull depression of the scene'. The journey is a secularized epic descent into the underworld on 'old Charon's boat, conveying melancholy shades to judgement'. Judgement of whom and for what is not specified, but the post-epic journey through the swamp as modern secular hell is part of the test which Martin must complete to achieve manhood on his journey to return home.

From this point the situation can only deteriorate as these are only the approaches to the underworld; this is not Hell itself but only the journey across the River Styx:

> as they proceeded further on their track, and came more and more towards their journey's end, the monotonous desolation of the scene increased to that degree, that for any redeeming feature it presented to their eyes, they might have entered, in the body, on the grim domains of Giant Despair. A flat morass, bestrewn with fallen timber; a marsh on which the good growth of the earth seemed to have been wrecked and cast away, that from its decomposing ashes vile and ugly things might rise; where the very trees took the aspect of huge weeds, begotten of the slime from which they sprung, seeking whom they might infect, came forth at night, in misty shapes, and creeping out upon the water, hunted like spectres until day; where even the blessed sun, shining down on festering elements of corruption and disease, became a horror; this was the realm of Hope through which they moved.

Giant Despair is a character straight out of Bunyan's *Pilgrim's Progress* who owns 'Doubting-Castle' (Bunyan, 2008, p. 116). Dickens transforms despair into the characteristic of capital 'D' Despair of gigantic proportions and transfers the giant's domain from a castle into a dismal swamp and slough of despond.

Although for Dickens Eden is ironically not a paradise but a festering sink in which the elements are improperly mixed and from this improper mixing spring hideous and monstrous deformities and mutants, by a further countervailing irony the wetland chaos, as I suggested in the first chapter, could be seen to be God's first work, not so much Eden, the pastoral paradise, but the pristine pre-Edenic, pre-pastoral and pre-paradisaical wetland. Dickens' Eden, however, is far from that watery chaos. It is both a 'hideous swamp' 'choked with slime' and a 'dreary situation' that produces 'dread' and 'the apprehension of death'. Like the approaches to the swamptown in which the elements of heat, water, air and earth are mixed, so too in Eden itself they are mixed where 'a fetid vapour, hot and sickening as the breath of an oven, rose up from the earth, and hung on everything around'.

At Eden Martin could not leave his footprints on 'the marshy ground' for as soon as his foot sank into it, 'a black ooze started forth to blot them out'. Martin is also unable to leave much of a mark on the land, or more precisely wetland, as it was 'mere forest' whose:

> trees had grown so thick and close that they shouldered one another out of their places, and the weakest, forced into shapes of strange distortion, languished like cripples. The best were stunted, form the pressure and want of room; and high about the stems of all, grew long rank grass, dank weeds, and frowsy underwood: not devisable into their separate kinds, but tangled

> all together in a heap; a jungle deep and dark, with neither earth nor water at its roots, but putrid matter, formed of the pulpy offal of the two, and of their own corruption.

This wetland is a nightmare for the taxonomizing botanist who would seek to separate out its specimens of flower, fruit and leaf into genus and species. This wetland is not just the excrement of earth and water, but the corrupt, decomposing excrement of earth and water. More generally, wetlands are a taxonomic anomaly in a classificatory order predicated on a hard and fast distinction between land and water, time and space.

Dreary impressions of solitude and decay

A telling instance of this figuring is found in Wilkie Collins' *The Woman in White* of 1859–60, set mainly in Blackwater Park with its Blackwater Lake. Marian Halcombe takes a walk in the Park and finds herself:

> standing suddenly on the margin of a vast open space, and looking down at [. . .] Blackwater lake [. . .] The lake itself had evidently once flowed to the spot on which I stood, and had been gradually wasted and dried up to less than a third of its former size. I saw its still, stagnant waters, a quarter of a mile away from me in the hollow, separated into pools and ponds by twining reeds and rushes, and little knolls of earth. On the farther bank from me the trees rose thickly again, and shut out the view, and cast their black shadows on the sluggish, shallow water. As I walked down to the lake, I saw that the ground on its farther side was damp and marshy, overgrown with rank grass and dismal willows. The water, which was clear enough on the open sandy side, where the sun shone, looked black and poisonous opposite to me, where it lay deeper under the shade of the spongy banks, and the rank overhanging thickets and tangled trees. The frogs were croaking, and the rats were slipping in and out of the shadowy water, like live shadows themselves, as I got nearer to the marshy side of the lake. I saw here, lying half in and half out of the water, the rotten wreck of an old overturned boat, with a sickly spot of sunlight glimmering through a gap in the trees on its dry surface, and a snake basking in the midst of the spot, fantastically coiled and treacherously still. Far and near the view suggested the same dreary impressions of solitude and decay, and the glorious brightness of the summer sky overhead seemed only to deepen and harden the gloom and barrenness of the wilderness on which it shone. I turned and retraced my steps to the high heathy ground [. . .].
>
> (Collins 1974, pp. 227–228)

This passage reproduces much of the traditional Western iconography and phenomenology of the wetland. The marshy side of the lake is figured as a kind of Eden-after-the-Fall in which a satanic serpent lurks. Dante and Milton had gone even further than Collins in satanizing the swamp as part of their theologizing of the landscape (as we have seen in earlier chapters): Dante by figuring one circle of hell as a slimy Stygian marsh; and Milton by figuring Satan as a draconian, monstrous swamp serpent who is generated out of the slime of hell.

What is probably more remarkable, though, about this passage from *The Woman in White* than its reproduction of traditional Christian iconography, and its moralization of the marsh, its *marécage moralisé*, is the trajectory it traces through space from the high heathy, even healthy, ground descending spatially and morally to the low poisonous, decidedly unhealthy wetland, and the juxtaposition it sets up between the sublime sky and the slimy swamp; between the openness of the former and the claustrophobia of the latter; the life and light of the former and the death, darkness, decay and disease of the latter; between the glories of the former and the wasteland of the latter.

The narrator/character of Marian Halcombe is positioned as first person in relation to the landscape, or more precisely what could be called the wetlandscape, or at least its visual space, entirely via the sense of sight. The repetitious insistence on, and of, 'I saw' places her in relation to various 'views,' or lack of them, the landscape reduced to occasions for taking, or thwarting, 'views'. She is not represented as hearing, touching, smelling or tasting the wetlandscape but is placed in a position of mastery from which to look down on the wetland in order to know it and so in order to master it, even though it is waste land. In fact, the 'I' is arguably constituted as bourgeois, indeed petit bourgeois, individual, albeit masculinized, subject insofar as the waste/wet land is postulated as alien other.

The figuring of the wetland as wasteland is exacerbated for Marian one windy and cloudy morning when 'the rapid alternations of the shadow and sunlight over the waste of the lake made the view look doubly wild, weird, and gloomy' (Collins, 1974, p. 253). The surface of the water reflects and 'heightens' the depths of gloom of the sky on a dark and overcast day, whereas on a bright and glorious day it is juxtaposed with it. The surface of water and the depths of the sky are set up with the petit bourgeois individual consciousness situated between, and mediating, them.

The wetland wasteland is quite useless from a capitalist agricultural point of view, though it may be picturesque from an aesthetic point of view and yet at the same time the perfect setting for a murder, and so associated with illegal death, three facts which are associated in Sir Percival's mind in *The Woman in White* when he relates how:

> 'Some people call that picturesque,' said Sir Percival, pointing over the wide prospect with his half-finished walking-stick. 'I call it a blot on a gentleman's property. In my great-grandfather's time the lake flowed to this place. Look at it now. It is not four feet deep anywhere, and it is all puddles and pools. I wish I could afford to drain it, and plant it all over. My bailiff (a superstitious idiot) says he is quite sure the lake has a curse on it, like the Dead Sea. What do you think, Fosco? It looks just the place for a murder, doesn't it?'
>
> (Collins 1974, p. 253)

For Sir Percival the wetland is a blot, a black mark, on a gentleman's property, on its clean and proper surface, on his character, on its title deeds and in its ledgers, and not even a debit in red. Sir Percival refers to this blot in the twin indexical gestures of the demonstrative 'that' (with no following substantive as if it were indescribable) and pointing with his stick as if he were unable to describe it adequately, or even get it into words.

Indeed, the wetland constitutes the unutterable in the Western tradition; hence the recourse that is invariably made, and Marian Halcombe as a narrator of *The Woman in White* is no exception, to the lexicon of what could be called standard swampspeak with its vocabulary of such well-worn words as 'dismal' (Collins, 1974, p. 279), 'dreary' (pp. 228, 253 and 280), 'desolate' (p. 280), 'gloomy' (p. 253) and so on. Wetlands have almost invariably been represented in the patriarchal Western tradition in metaphors of despair and despondency in an overworking of the lower echelons of the empathetic fallacy and of the nether regions of the psychopathological register.

Great Grimpen Mire

The slough of despond is the ideal setting or backdrop for murder and for the private detective story as it is a place of almost impenetrable evil and darkness which can only be pierced by the penetrating light of reason brought to bear upon it by the superior intellect and insight of the great man himself. It is hardly surprising then that the most famous private detective of them all, in probably his most famous story of all, should have to deal with a slough of despond. This is, of course, Sherlock Holmes in *The Hound of the Baskervilles* with its great Grimpen Mire.

When Stapleton takes Dr Watson on a tour of the moor, he asks Watson if he notices '"those bright green spots scattered thickly over it"'. Watson replies, 'Yes, they seem more fertile than the rest,' to which his guide laughs and goes on to say, 'That is the great Grimpen Mire,' which he describes as 'the bog-hole' and as 'an awful place'. By acknowledging the fertility of the mire, Watson was a wetlands ecologist before his time! While the Mire is being discussed 'a long, low moan, indescribably sad swept over the moor. It filled the whole air, and yet it was impossible to say whence it came. From a dull murmur it swelled into a deep roar, and then sank back into a melancholy, throbbing murmur once again.' Watson asks what the sound is only to be told that '"the peasants say it is the Hound of the Baskervilles calling for its prey"'.

Watson is not satisfied with this answer, dismissing it as 'nonsense' and pressing Stapleton for a more rational explanation, which he feebly offers:

> 'bogs make queer noises sometimes. It's the mud settling, or the water rising, or something.' 'No, no that was a living voice.' 'Well, perhaps it was. Did you ever hear a bittern booming?' 'No, I never did.' 'It's a very rare bird – practically extinct – in England now, but all things are possible upon the moor.' 'It's the weirdest, strangest thing that ever I heard in my life.' 'Yes, it's rather an uncanny place altogether.'

Indeed, the wetland is the uncanny place *par excellence* for Western culture (see Giblett, 1996, pp. 30–34; 2009, p. 21).

Watson is greatly affected by 'the melancholy of the moor' and reports to Holmes from what he calls:

> this most God-forsaken corner of the world. The longer one stays here the more does the spirit of the moor sink into one's soul, its vastness, and also its grim charm. When once you are out upon its bosom you have left all traces of modern England behind you.

This conclusion is hardly surprising because once you have entered the mire mother, or the mother of all mires, you have entered a pre-modern, even pre-Christian, landscape, or more precisely wetlandscape, which is the precursor of the modern landscape and the 'foundation' for it. Even though one may avoid sinking into it physically, it sinks into one metaphysically and psychologically. It is not a settled and stable object, but its own free agent, a dynamic process which cannot be kept within strict bounds, whose primary horror is that it breaks boundaries.

In the conclusion to the story Holmes and Watson follow Stapleton into 'the widespread bog' with its 'green-scummed pits and foul quagmires' where:

> rank weeds and lush, slimy water-plants sent an odour of decay and a heavy miasmatic odour onto our faces, while a false step plunged us more than once thigh-deep into the dark, quivering mire [. . .] Its tenacious grip plucked at our heels as we walked, and when we sank into it it was if some malignant hand was tugging us down into those obscene depths, so grim and purposeful was the clutch in which it held us.

Although Watson as a modern medical man does not seem to subscribe to a miasmatic theory of malaria, he supports a miasmatic theory of melancholia. He also anthropomorphizes the mire by giving it hands to pull Holmes and him down into its clutching, (s)mothering embrace as it had pulled Stapleton 'down in the foul slime of the huge morass'.

The valley of the shadow of death

Similarly, Fergus Hume in his best-selling detective novel and 'the first popular Melbourne novel', according to John Arnold (1983, pp. 7–8), *The Mystery of a Hansom Cab*, first published in 1886, evokes vividly the Dantesque 'Infernal Regions' 'off Little Bourke St'. Like the epic hero descending into the underworld to slay monsters and return home triumphant, 'the detective led the way down a dark lane, which felt like a furnace owing to the heat of the night' (Hume, 1999, pp. 178–179). The infernal distress of the slums of Bourke Street can be contrasted with what Marjorie Clark (in Arnold, 1983, p. 69) in 1927 described as 'the cool charm of its [Collins Street's] arcades'. Worse for the detective entering the slums of Bourke Street, 'it was like walking in the valley of the shadow of death [. . .] And, indeed, it was not unlike the description in Bunyan's famous allegory what with the semidarkness, the wild lights and shadows' (Hume, 1999, p. 180). Hume combines Dante and Bunyan (and the biblical Psalmist, as Bunyan is drawing on *Psalm* 23) in an omnibus and ominous grab bag of tropes for the urban underworld.

As an aside here, these pages are excerpted and reproduced as a quintessential depiction of the 'mean streets and back alleys' of late nineteenth-century Melbourne as depicted in crime fiction for the anthology *Literary Melbourne* edited by Stephen Grimwade (2009, pp. 187–190). The reference to 'the valley of the shadow of death' is later misquoted as 'the shadow of the valley of death' (Grimwade, 2009, p. 244). This misquotation gives a new nuance to the idea of death valley transposed here from the remote deserts of California to inner-city Melbourne. Both are hot and deadly places. As Michael Cannon (1976, pp. 42 and 43) puts it, 'the Angel of Death came early and stayed late in the Melbourne of 1892 and 1893' with 'epidemics of influenza, typhoid and measles [. . .] which [. . .] killed thousands'. It is only a hop, skip and a jump from Bunyan's and the Psalmist's 'valley of the shadow of death' to Hume's alley of the shadow of death and Grimwade's shadow of the valley of death.

The detective's journey into the urban underworld is not only physical but also moral and allegorical. The detective novel is a secular allegory of literal and spiritual descent and ascent, of degradation and salvation, just like the sacred allegory of Bunyan's *Pilgrims Progress*. In the case of Sherlock Holmes, the detective's journey also includes a descent into the Bunyanesque 'Slough of Despond' of great Grimpen Mire in *The Hound of the Baskervilles* (as we have already seen). Similarly, Flinders Street in 1845 was described as 'that slough of despond' (cited by Annear, 2014, p. 40). The swamp outside the city in the country, or inside the city in the muddy streets or slums, is a secularized satanic space. Perhaps it is fitting that the *flâneur* is a creature of the arcades, which Johann Geist (1983, p. vii) describes in his monumental history of the arcade as 'a secularised sacred space'. The swamp is a secularized satanic space for the detective that is a sacred space and place for indigenous, traditional peoples. The secularized satanic space of the swamp could be the native or natural swamp in the country, or the feral or cultural 'swamp' of the slums in the city in which the latter is used as a figure for the latter.

In *Pilgrim's Progress* Bunyan (2008, p. 65) takes the fourth verse of *Psalm* 23 referring to the valley of the shadow of death for his text and sermonizes on how 'Christian must needs go through it, because the way to the Celestial City lay through the midst of it'. Typically, the detective in the modern city must go through the valley of the shadow of death of the underworld of the slums in order to protect the upper city of the upper ten thousand living in the upper world of the crystalline celestial city. Or the hero of the modern epic of the novel, such as Jean Valjean in Hugo's *Les Miserables*, must descend into the underworld of the sewers of Paris (see Giblett, 2016, chapter 3). The modern hero of the tourist can even follow in Valjean's footsteps and go on a tour of the sewers of Paris in the marshy underworld of 'Lutetia', the Latin-cum-Celtic name for the 'filthy marsh' in which the city was founded. Although Melbourne as the 'Paris of the South' does not have such an official tour of its sewers, 'a small group has for several decades mounted regular unlawful explorations of Melbourne's stormwater drains and tunnels' in what Kristin Otto (2005, p. 175) calls 'a Yarra [River] underworld'. Sophie Cunningham (2011, p. 147) also comments that 'while Melbourne does not boast Paris's hundreds of kilometres of underground tunnels, it does have Anzac, a cavernous drain under South Yarra where many parties have been held over the years'.

Later in *Pilgrim's Progress* Christian is informed that 'the valley [of the shadow of death] itself' is:

> as dark as pitch: we also saw there the hobgoblins, satyrs, and dragons of the pit; we heard also in that valley a continual howling and yelling, as of a people under unutterable misery, who there sat bound in affliction and irons; and over that hung the discouraging clouds of confusion; Death also does always spread his wings over it. In a word, it is every whit dreadful, being utterly without order.
>
> (Bunyan, 2008, pp. 67 and 69)

As the underworld is a monstrous place itself inhabited by monsters, such as dragons, so is 'the monster city'. In similar vein, Little Bourke Street in *The Mystery of a Hansom Cab* has its 'weird and grotesquely horrible' inhabitants (Hume, 1999, p. 220). In a word, they are monstrous. Little Bourke Street contrasts with Bourke Street proper, not only in the type of inhabitants, but also in the mode of illumination. Bourke Street is described 'as the brilliantly lit street' produced by 'electric lights', which highlight the members of the crowd caught in 'the full glare of the electric light'. By contrast, Little Bourke Street is lit by 'sparsely scattered gas lamps with their 'dim light', whereas the lanes have no lights or lamps so they are dark, or:

> not quite dark, for the atmosphere had that luminous kind of haze so observable in Australian twilights, and the weird light was just sufficient to make the darkness visible.
>
> (Hume, 1999, pp. 178–180)

Typically, the fictive detective of the modern city enlightens the benighted, brings light to darkness and illuminates the crepuscular gloom of crime and grime.

In her biographical and critical study of Hume's novel, Lucy Sussex (2015, p. 7) argues that it had 'an important role in establishing detective fiction as a publishing category' as he is 'one of the most influential crime writers of all time' who wrote 'the biggest-selling crime novel of the nineteenth century, and one of the most important Australian books ever'. Sussex traces the etymology of the word 'detective' and how it literally means 'de-roofing' as 'a detective raises the roof, figuratively'. The detective in the detective story, and the detective storyteller, raises the roof of dwelling spaces, looks inside and reveals what is inside to the reader. The detective and detective story reader are positioned as snooping voyeurs.

Sussex (2015, pp. 9–10) also traces how this uncovering and de-roofing had a demonic function and cites Dickens in *Dombey and Son* published in 1848 who pleaded for 'a good spirit who would take the house-tops off'. David Grann (2017, p. 57) similarly traces how 'the term "to detect" derived from the Latin verb to "unroof," and because the devil, according to legend, allowed his henchmen to peer voyeuristically into houses by removing their roofs, detectives were known as "the devil's disciples". In the Sherlock Holmes story 'A Case of Identity', Conan Doyle has Holmes describe the detective's work of unroofing in similar terms (albeit without the etymology and the theological overtones). The detective in the detective story, and the detective storyteller too, perform these socially useful, but morally ambiguous, roles of the good, or demonic, spirit. Both were *flâneurs* who entered 'the city's central hell [of] the slums', as Sussex (2015, p. 60) puts it in relation to Hume and *Hansom Cab.* As with the circles of Dante's hell, or Inferno, the city has a centre, and its centre is the lower depth of its slums and sewers, the grotesque lower urban strata. The seemingly most secularized literary genre of the detective story about the modern city has strong theological undertones.

A fountain of writhing and reptilian life

C. S. Lewis explains in the 'Afterword to the Third Edition' of *The Pilgrim's Regress: An Allegorical Apology for Christianity, Reason, and Romanticism* (an unlikely and incompatible grab bag) that 'the dominant image of my allegory [is] the barren, aching rocks of its "North," the fetid swamps of its "South," and between them the Road on which alone mankind can safely walk' (2014, p. 238). Beside the road, Pilgrim John sees 'a concourse of living creatures' (p. 216). When John gets closer, he sees that 'the concourse was of men, but they lay about in such attitudes and were so disfigured that he had not recognized them for men' (p. 217). These men are monstrous creatures of the marsh. They live on the swampy south side of the road where 'the ground was very soft' (p. 217). They live in a marsh where the ground was not only very soft, but also very wet and reedy as 'some of them were half under water and some hidden in the reeds' (p. 217). Their monstrosity consists of a hideous, syphilitic disfigurement as 'all seemed to be suffering from some disease of a crumbling and disintegrating kind. It was doubtful whether all the life that pulsated in their bodies was their own' (p. 217). Their life is the diseased life of the marsh, and the disease itself.

The disease is reptilian as John sees that 'the whole assembly was but a fountain of writhing and reptilian life [. . .] sprouting out of human forms' and 'a fountain of vermin' (p. 217). The guide advises John not to 'lag [. . .] [as] this is a very dangerous place [. . .] This place is called Luxuria' (p. 217), one of the seven deadly sins of lust and lechery. The name of the place, and the chapter, leaves the reader in no doubt that the men are suffering from a sexually transmitted disease. It seems that Lewis may have been imagining and invoking the monstrosity of both Pieter Bruegel the Elder's late medieval depiction of 'Luxuria' of 1558 and the modern syphilis-sufferer's hospital ward.

A witch administers a cup of unnamed liquid to the men. One of the men, whose fingers are unusually supple at the joints, eventually sinks down 'in the swamp with a groan. And the worms where there should have been fingers were unmistakable' (p. 218). He has become vermin, a creature of the wet ground, a worm in and of the coldness of the grave. Worms in biblical terms, as Mary Douglas (1966, p. 56) puts it, 'belong in the realm of the grave with death and chaos'. They are a throwback to the primeval chaos of the first day of creation in which the world was wetland and before God divided dry land from wet land and water. Instead of the wetland being a place of new life and goodness as in modern ecology, in Lewis's allegorical theology it is a place of decomposition, death and evil, including witches. John dreams that 'the witch came to him walking softly in the marshy ground by the roadside' (Lewis, 2014, pp. 218–219). Lewis' witch is not a good wise woman administering herbal remedies collected from plants growing in the wetlands along the side of the road, but the precise opposite.

Either side of the road is the home of two dragons, 'the Northern (or cold) dragon' and 'the Southern (or hot) dragon' (Lewis, 2014, pp. 221–227). The northern dragon is an earth dragon, like the dragon of the poem *Beowulf*, as the north is a place of 'barren, aching rocks'. The southern dragon is a swamp dragon, like the dragon of the legend St George, as the south is a place of 'fetid swamps'. Like Stapleton advising Watson in *The Hound of the Baskervilles*, 'the guide' advises John to 'go where it is least green [. . .] for there is the ground firmest' (p. 225). Green signifies fertility, soft ground and 'bog holes' (p. 226), just as in great Grimpen Mire. Vertue is accompanying John and they head southward 'feeling out the fen-paths' (p. 221), like Watson and Holmes in great Grimpen Mire.

Only one dragon has survived, as one dragon has eaten the other, 'otherwise he would not be a dragon', the guide tells the pilgrim. Both the guide and John are well-read in Renaissance English literature, such as Francis Bacon's early seventeenth-century essay, 'Of Fortune'. John supposes that there is only one dragon and the guide confirms this supposition by citing in part Bacon's Latin maxim: '*serpens nisi serpentem comederit –*' (Lewis, 2014, p. 221). The guide does not complete the maxim as he knows that John will able to complete it, and understand it. Both the guide and the pilgrim were well-educated in the English 'public' (private) school system of the early twentieth century when Lewis was growing up, in which Latin and the English classics were mandatory parts of the curriculum (unlike for many of Lewis's readers). Indeed, Lewis may have read the edition of Bacon's essays edited by J. A Hunter who was 'one of the National Society's Examiners of Middle-Class Schools', as his byline on the title page of this late nineteenth century edition proclaimed. The rest of the Latin maxim is: '*non fit draco'* (Bacon, 1873, p. 163). The whole maxim can be translated as: 'a serpent must have eaten another serpent before he can become a dragon'. Serpents are cannibalistic and potential dragons. Serpents only reach their full potential of becoming dragons by consuming other serpents. They embody the horror of cannibalism. They are monsters who deserve to be killed, so John duly obliges.

Like Christian in Bunyan's *Pilgrim's Progress*, John in Lewis's *Pilgrim's Regress* is a sword-wielder; unlike Christian, though, but like both Beowulf and St George, John is a successful dragon-slayer. Like St George, but unlike Beowulf, John survives his battle with the dragon, when 'with a jab of his sword into the under-side of the brute, it went in to the hilt, but there was no blood'. The dragon is a bloodless creature, hardly living at all. He lives a living death, like the disease-ridden men. They are manifestations of the dragon, and the dragon is a manifestation of the disease. The disease-ridden young man with fingers like worms is becoming-dragon as 'wyrme' is the Old English name for a serpent and a dragon (as we saw in a previous chapter). The dragon is not really a living, breathing dragon at all, certainly in some respects not like any previous dragon.

Indeed, he is more metallic than organic. The inside of his mouth is 'not red [. . .], but grey like lead' (Lewis, 2014, p. 225). Unlike the fiery breath of every previous dragon, 'the breath of the creature was freezing cold'. Rather than sapping John of energy, the dragon's cold breath invigorates him and so John makes 'thrust after thrust into the brute's throat' until he had 'a dead reptile at his feet' (p. 225). The dead dragon either reverts to serpent, or the cannibalistic dragon, who had been a serpent who ate another serpent to become a super-serpentine (and still a reptilian) dragon, remains a reptile when he dies. Bacon did not envisage such a reverse transformation or continued reptilian existence when the dragon dies, nor do John and the guide comment on it.

In Lewis's *The Pilgrim's Regress* Luxuria, or the swamp of lechery and marsh of monstrosity, functions like the Slough of Despond in Bunyan's *Pilgrims Progress* as a sump of iniquity and cesspool of evil. Similarly, the dragon in Lewis's allegory functions like Apollyon in Bunyan's as an embodiment, or more precisely montrosification, of evil to avoid or to kill in order to destroy evil. The disease-ridden and monstrous men, the cold, metallic and serpentine dragon, and their marshy habitats all function as the valley of the shadow of death in Lewis's *Pilgrim's Regress* as in Bunyan's *Pilgrims Progress.* Death is bodied forth and adumbrated in the syphilitic men and the serpentine dragon.

The parasite-infested swamps of nowhere

The American cultural studies academic Donna Haraway uses the figure of the slough of despond with pejorative overtones (though not, of course, as an allegorical emblem for sin) in her proclamation that, 'like Christian in *Pilgrim's Progress* [. . .], I am committed to skirting the Slough of Despond and the parasite-infested swamps of nowhere to reach more salubrious environs' (Haraway, 1992, p. 295; see p. 329, n.1 for her exclusion of the allegory of sin as Slough of Despond in invoking *Pilgrim's Progress*). One wonders salubrious for whom? In whose terms? And if the Slough of Despond is not an allegorical emblem for sin, what is 'sin' in Haraway's theologized, or at least moralized, wetlandscape? What is 'nowhere'? Is it 'the amniotic effluvia of terminal industrialism' that Haraway refers to later (but without making any explicit connection)? If so, then the figure would have a critical edge and contemporary pertinence; if not, then Haraway's use of the figure would seem to be gratuitously uncritical about the use of tropes and blind to the politics of place (placist, and even misaquaterrist).

Yet she argues in these very same two pages that 'nature is [. . .] a *topos*, a place', that 'nature is also *trópos*, a trope' and later for a 'politics of articulation' which speaks with an intersubject rather than for 'a politics of representation' which speaks for and on behalf of an object (Haraway, 1992, pp. 296 and 311–313). But as nature is also place*s, topoi*, as different *topoi* are troped in different ways with some being valorized at the expense or to the detriment of others, so there is a politics of the articulation of tropes of topes, a politics of *trópoi* of *topoi*, not least of wetlands. Haraway cares about 'the survival of jaguars and the chimpanzee, and the Hawaiian land snails, and the spotted owl, and a lot of other earth*lings*' (Haraway, 1992 p. 311; my emphasis), but does she care about the survival of sloughs (of despond), (parasite-infested) swamps and uncharismatic microfauna (like parasites)? Unlike Christian in *Pilgrim's Progress* and Haraway in her essay, and like Henry David Thoreau (as we will see in chapter 8), I am a committed pilgrim striking out into sloughs of hope and into richly fertile swamps everywhere sinking into their salubrious environs.

References

Annear, R. 2014. *Bearbrass: Imagining Early Melbourne*. Melbourne: Black Inc.

Arnold, J. 1983. *The Imagined City: Melbourne in the Mind of its Writers*. North Sydney: George Allen and Unwin.

Bacon, F. 1873. Of Fortune. *In*: J. Hunter, ed., *The Essays of Lord Bacon*. London: Longman, Green, pp. 163–165. First published 1612.

Bunyan, J. 2008. *The Pilgrim's Progress*. R. Pooley, ed. London: Penguin. First published 1678.

Burton, R. 1932. *The Anatomy of Melancholy*. H. Jackson, ed. London: Dent. First published 1638.

Cannon, M. 1976. *The Land Boomers*. New Illustrated Edition. Melbourne: Thomas Nelson.

Collins, W. 1974. *The Woman in White*. Harmondsworth: Penguin. First published 1859–1860.

Cunningham, S. 2011. *Melbourne*. Sydney: New South.

Douglas, M. 1966. *Purity and Danger: An Analysis of the Concepts of Pollution and Taboo*. London: Routledge.

Eagleton, T. 1986. *William Shakespeare*. Oxford: Basil Blackwell.

Fritzell, P. 1978. American Wetlands as Cultural Symbol: Places of Wetlands in American Culture. *In*: P. Greeson, J. Clark and J. Clark, eds, *Wetland Functions and Values: The State of Our Understanding*, Minneapolis: American Water Resources Association, pp. 523–534.

Geist, J. 1983. *Arcades: The History of a Building Type*. The Massachusetts Institute of Technology.

Giblett, R. 1996. *Postmodern Wetlands: Culture, History, Ecology*. Edinburgh: Edinburgh University Press

Giblett, R. 2016. *Cities and Wetlands: The Return of the Repressed in Nature and Culture*. London: Bloomsbury.

Grann, D. 2017. *Killers of the Flower Moon: Oil, Money, Murder and the Birth of the FBI*. London: Simon and Schuster.
Grimwade, S. ed. 2009. *Literary Melbourne: A Celebration of Writing and Ideas*. Prahran: Hardie Grant.
Haraway, D. 1992. The Promises of Monsters: A Regenerative Politics for Inappropriate/d Others. *In*: L. Grossberg, C. Nelson and P. Treichler, eds, *Cultural Studies*, New York: Routledge, pp. 295–337.
Hume, F. 1999. *The Mystery of a Hansom Cab*. Melbourne: Text. First published 1886.
Huxley, T. H. 1989. Struggle for Existence. *In*: P. Kropotkin, *Mutual Aid: A Factor of Evolution*, Montreal: Black Rose. First published 1888.
Kant, I. 1952. *The Critique of Judgement*. J. Meredith, trans. Oxford: Clarendon.
Kant, I. 1960. *Observations on the Feeling of the Beautiful and Sublime*. J. Goldthwait, trans. Berkeley: University of California Press.
Lewis, C. S. 2014. *The Pilgrim's Regress: An Allegorical Apology for Christianity, Reason, and Romanticism*. Grand Rapids, MI: William B. Eerdmans. First published 1933.
Otto, K. 2005. *Yarra: A Diverting History*, Melbourne: Text.
Sussex, L. 2015. *Blockbuster! Fergus Hume and the Mystery of a Hansom Cab*. Melbourne: Text Publishing.
Tillyard, E. M. W. 1963. *The Elizabethan World Picture*. Harmondsworth: Penguin. First published 1943.

7 God's and nature's nation

John 'the Baptist' Muir and US national parks

National parks are commonly regarded today as conservation landscapes, but the initial impetus to set them aside was monumentalist and sanctuarist, rather than conservationist and sacral. The monument commemorated a significant moment in time and the sanctuary set aside a special place in space. They were products of the patriarchal paradigm. By setting aside monumental, usually mountainous and forested, landscapes, national parks were the inheritors and perpetrators of a romantic landscape aesthetic, especially of the sublime. The first national parks in the United States had spectacular natural features, such as peaks, canyons, waterfalls or geysers or various combinations thereof. The first national parks were parklands, not parkwaters; drylands, not wetlands. It took sixty years for a national park to be declared in a wetland.

These national parklandscapes were imbued with nationalist significance. The United States sought, found and maintained a sense of national identity in privileged national landscapes. The impetus to create national parks was nationalist, as they legitimated the nation's claim to the national territory. National parks nationalized special places by carving out and enclosing national space within global space. The national parks also nationalized time by demarcating and inscribing national history within global history. The national parks were a kind of nationalist graffiti marked in rocky landscapes. They made their mark in time and space. The ruinous landscapes of national parks were often likened to cathedrals. They gave a sense of history and longevity to a new nation. By nationalizing time and space national parks dispossessed indigenous peoples from their lands and from their past and future.

This chapter deconstructs the monumentalism and sanctuarism in the writings of John Muir, father of the American national parks, or at least of the idea for them. It also critiques the gender politics of Muir's contradictory ideas of nature divided between a stern and sublime Father God in the mountains and a soft and smiling Mother Goddess in the swamps. Although early in his career Muir had a close encounter with the latter in a Canadian wetland which converted him to nature-worship (see Giblett, 2015, pp. 155–159), for the remainder of his life he preferred and privileged the former over the latter. Although nature for Muir was imbued with a romantic, aesthetic theology, it was always alive, never mere dead matter. Muir was both sanctuarist in his views about national parks and sacralist in his views about nature (see Giblett, 2011, chapter 1).

John Muir was, and still is, a monumental figure in the American conservation movement, if not its patriarch. He has been described as 'the spiritual father of American environmentalism' (Worster, 1993, p. 189), as 'the father of our [United States'] national parks and forest reservations' (Tilden, 1970, p. 537), and as 'a patriarch of the National Park system' (Mitchell, 1994, p. 52). The father of the idea for national parks is generally recognized as George Catlin who is credited with proposing it in 1832 (see Giblett, 2011, chapter 8). In the present chapter I consider Muir's ideas about national parks and the gender politics of his paradoxical views of nature. For Muir nature in general is sacralized, yet his views on where and how nature is sacralized depend on the aspect or site under discussion. For Muir nature in mountains and forests was aestheticized and sublimated, especially if such areas were 'set aside' and preserved in the sanctuaries or 'modern cathedrals' of national parks, whereas nature for him could also be inhabited, invigorated and enlivened by all living things, including rock and stone, and even human beings (but not indigenous peoples).

Under Muir's leadership the American conservation movement has largely been concerned with conserving sanctuaries, those special places for plants and animals and those aesthetic landscapes for taking refuge from the depredations of modern urban industrial capitalism. American conservationism has often been more precisely 'sanctuarism', concerned with giving sanctuary and preserving sanctuaries rather than with caring for the ecosphere as a whole. Certainly those special places need to be conserved – humans need sanctuaries and sanctuaries are needed for plants and animals. In the famous John Muir–Gifford Pinchot struggle over the damming of Hetch Hetchy valley in Yosemite National Park and its descendants I am on Muir's and his children's side every time.

The American conservation movement arose in the aftermath of what Stephanie Coontz (cited by Flichy 1995, p. 67) calls 'the apogee of the private sphere' (1870–1890) in which 'refuge' and 'sanctuary' were repeatedly and routinely applied to the home. By taking these terms from the private sphere and applying them to the biosphere, the latter was privatized. Even public lands were construed in private terms. The commons, owned by none and shared by all, were colonized in public lands such as national parks and enclosed in private terms such as 'refuge' and 'sanctuary'.

This care for special places needs to be extended to the earth household as a whole, as a special space with special places. Deconstructing Muir's work with its romantic aesthetic of the sublime as a naturalistic theology and valuing his work on nature as living being is one step to doing so. It is also part of a proactive political ecology that promotes and exercises earthly mutuality – and not a reactionary sanctuarism constantly trying to save endangered species or special places threatened with destruction. Rather than always fighting rearguard actions, political ecologists are the avant-garde.

Muir sublimated mountains and forests into a patriarchal God not so much in what Murray Bookchin (1991, p. il; for Muir's Calvinist upbringing see Fox, 1981, pp. 28–30) calls his 'inverted Calvinism' (as the relationship between God and nature is not merely turned upside down) as his naturalistic Calvinism, or perhaps more precisely naturalistic Campbellism.[1] Instead of 'the natural world as a path to a knowledge of God' (Grove 1995, p. 15) as in Calvinism, the natural world for Muir was a manifestation of God. Although he is often regarded as a pantheist, Michael Zimmerman (1994, p. 385, n.48) argues that 'Muir may have been a *panentheist*, one who regards nature as a manifestation of but not exhaustive of the divine'. Muir's God was incarnated in His son of the elected, and even predestined,[2] elements and features of mountains and forests such as geysers and glaciers especially, but also in rock and stone, wood and water, plants and animals, and even human beings (though again not indigenous peoples).

Yet when Muir celebrates living nature not just in forests and mountains, but also in glaciers and swamps, at work destroying and creating life, nature is thereby enlivened or invigorated in 'panentheism' (see Sheldrake, 1991, p. 198). Muir here sacralizes nature as 'Mother Earth' or 'Mother Nature'.[3] Muir vacillates between nature aestheticized (especially sublimated and incarnated) for nature lovers, nature preserved (especially in the sanctuaries of the modern cathedrals of national parks) for city dwellers and tourists, and nature inhabited (especially enlivened or invigorated by all things including human visitors to national parks, but not for Muir by indigenous human inhabitants or native title holders of national parks).

Cathedrals of the Modern World

In a series of articles published in *The Atlantic Monthly* during the 1890s and later collected in his book, *Our National Parks*, first published in 1901, Muir expressed his ideas and views about national parks in particular and nature in general. Muir also wrote specifically and lovingly of the first two national parks in the United States (and the world), Yellowstone National Park and Yosemite National Park. For Muir national parks, or more precisely forested and/or mountainous national parks, were seen as fulfilling an important inspirational role as modern cathedrals. Yet forested and/or mountainous national parks are more precisely America's outdoor modern cathedrals distinct, as we will see below, from the indoor modern cathedrals of the cinemas of industrial capitalism and the museums of the nation-state.[4]

Muir (1980, p. 127), however, goes one step further than simply regarding Yosemite National Park as an outdoor cathedral as he found 'the mighty wilderness of mountains' of the High Sierra not only like 'some Gothic cathedral' but also 'more abundantly spired than Milan's'. The inspiring High Sierran cathedral 'outspired' the smaller, less abundantly spired Christian cathedral. Muir was not alone among his conservationist predecessors and his contemporaries in this view of forested and mountainous regions. In 1858 Henry David Thoreau (1962, *XI*, p. 353) described in his *Journal* 'the quiet and somewhat sombre aisles of a forest cathedral'. Yosemite was regarded widely as 'a natural cathedral' in the nineteenth century (see Sears 1989, pp. 122 and 138). Indeed, one of the mountains of Yosemite is named Cathedral Peak, which Muir (1987, p. 198) described as 'a majestic temple [. . .] adorned with spires and pinnacles in regular cathedral style'.

Yet Yosemite was not a Christian cathedral. For Muir (1987, pp. 146 and 198) it was 'one of Nature's cathedrals, hewn from the living rock' and Cathedral Peak was 'hewn from the living rock'. It was left *in situ*, unlike the cathedrals of Europe built of dead stone, the extracted and hewn resource, the transported and erected object. Nature for Muir was living agent, not dead matter. No Sierra landscape that Muir (1987, p. 157) saw 'holds anything dead or dull, or any trace of what in manufactories is called rubbish or waste'. Mountainous landscape may have been wasteland in the eyes of his contemporaries, but for Muir (1987, p. 60) it was where 'God himself seems to be always doing his best here, working like a man in a glow of enthusiasm'. God's Promethean, thermodynamic and industrial workshop, and home, for Muir is up in the mountains, not down on the lowlands in the cities, the home of citizens, and certainly not even further down in the netherlands of the swamps, the home of the matrifocal Great Goddess (see Giblett 1996, especially chapter 6). As Muir (1987, p. 205) refers later to 'Nature's glacial workshops', God is Nature and Nature God for Muir.

Despite, or perhaps because of being God's best work, mountainous landscape did not fall into the conventional aesthetic category and experience of the picturesque. For William Gilpin (1794, p. 43) in the late eighteenth century, 'the spiry pinnacles of the mountain, and the castle-like arrangement of the rock, give no peculiar pleasure to the picturesque eye'. Similarly, a hundred years later Muir (1980, p. 104) claimed that 'to artists, few portions of the High Sierra are, strictly speaking, picturesque. The whole massive uplift of the range is one great picture, not clearly divisible into smaller ones'. In other words, the High Sierra with its phallic 'massive uplift' invokes the experience of the sublime as do 'the ineffably chaste and spiritual heights of the Fairweather Range [. . .] the whole making a picture of icy wildness unspeakably pure and sublime' (Muir, 1980, p. 14).

Muir enacts what Belden Lane (1988, p. 22) calls 'the sublime as the monumental', or perhaps more precisely, the sublime as the *experience* (the sublime is a *state* of aesthetic sensibility) of the *mountainous* (as distinct from the urban or built) monumental.[5] Again Muir was not alone in the nineteenth century in valorizing the mountainous as evocative of the sublime. Indeed, it was a commonplace of nineteenth-century landscape aesthetics (for example, Wordsworth; see Giblett, 2011, chapter 3). For Thoreau (1980, p. 44) too, 'sublimity and grandeur [. . .] belong to mountain scenery' and for John Ruskin (cited by Schama 1995, p. 513) also 'we take our idea of fearfulness and sublimity from the mountains and the sea', though he prefers the former to the latter.

Yet Yosemite was 'that sublime Sierra temple' not only because it was mountainous but also because it was forested. Muir (1991, p. 60; see also p. 226) describes how 'the sunbeams streaming through their [the pines'] feathery arches brighten the ground, and you walk beneath the radiant ceiling in devout subdued mood, as if you were in a grand cathedral with mellow light sifting through colored windows, while the flowery pillared aisles open enchanting vistas in every direction'.[6] For Muir, 'reservations of scenery', as for his compatriot and near contemporary Charles Eliot (cited by Thomas 1984, p. 269) writing in 1896, were 'the cathedrals of the modern world'. Or perhaps more precisely, *mountainous forest* reservations (the forerunners of national parks) of 'scenery', the habitual visual aesthetic of the sightseer and tourist, were modern cathedrals.

Yet the cathedral is very much an alien and imported construct imposed on an indigenous inhabitation of the land, the Christian (or in Muir's case his naturalistic Campbellism (but what of his panentheism here?)) proselytizing the 'pagan' and sublimating the chthonic. Couple the cathedral with the sublime in the monumental mountains and then set it aside in the sanctuary of the national park and the result is a neoromantic sacred sight (not site). In Muir's (1991, p. 69) naturalistic Campbellism a stern and craggy, patriarchal and mountainous triune God 'brood[s like the Holy Spirit] outspread over the predestined landscape' of the glacial mountains and valleys, instead of a stern and craggy, patriarchal and monumental God brooding spiritlike over predestined saved souls of Calvinism or over the primordial wetlandscape of the littoral, lacustrine and lagoonine as in the biblical creation story (as seen in the rereading of *Genesis* 1:2 in these terms in chapter one above).

By referring to 'the wombs of the ancient glaciers' Muir (1980, p. 222) displaces the beginnings of the world from the littoral to the glacial, from the world below to the world above. His patriarchal appropriation of maternal functions by birthing metaphors and his displacement of the primordial landscape from the littoral to the glacial which does not acknowledge the primordial womb of the wetland are both crucial for his sublimatory nature aesthetic and religion. In Muir's naturalistic Campbellism, the patriarchal God-function is sublimated into the monumental mountains where God the Father Law gives birth to new life with His own labour alone in the simulated womb of His glacial workshop. The matrifocal Great Goddess of the swamps is displaced upwards into patriarchally simulated into what Muir (1980, pp. 102 and 46) calls 'glacier wombs' and 'the broad white bosoms of the glaciers'. The glacial white bosom is a kind of *tabula rasa*, or in Freud's terms a mystic writing pad, on which Muir could inscribe his writings and in Freud's terms a screen on which he could project his phantasies, his daydreams. The whiteness of the bosom serves to smother any trace of the living black waters of the Great Goddess of the swamps, and to sanitize any messiness in parturition.

The forested and/or mountainous national park as modern cathedral is the outdoor extra-urban counterpart to the indoor urban cathedrals of the modern world. These are what David Day (1996, p. 182) calls 'the new stone-built cathedrals of knowledge, the public libraries, art galleries and museums' and what Paul Virilio (1989, p. 38) calls 'the cinema-cathedrals of the modern state'. Or perhaps more precisely, the cinema-cathedrals of the modern imperialist nation-state in hegemonic alliance with industrial capitalism. As with film, which was for Walter Benjamin (1986, p. 55) 'one of the most advanced machines for the imperialist domination of the masses', the national park is a machine for the imperialist domination of the earth. 'The empire of nature', as John MacKenzie (1988) calls it, is applicable not just to the nineteenth century in the colonies, but to the capitalist commodification of nature today in national parks, zoos, and other nature theme parks, as well as in nature documentaries, so-called 'natural products' (an oxymoron), and tourism.

Yet the national park and the cinema as modern, imperialist cathedrals performed different quasi-religious functions. The early cinemas, Virilio (1989, p. 31) argues, were regarded as 'deconsecrated sanctuaries in which, as Paul Morand put it, the public sensed the end of the world in an ambience of profanation and black masses'. For Muir, by contrast, Yosemite and Yellowstone National Parks were reconsecrated sanctuaries in which 'the public' could sense the evolutionary, albeit glacial, beginning, and even ongoing recreation, of the world in an ambience of reverence and sublime 'white' masses (unlike the mephitic and miasmatic atmosphere of the slimy, living black morasses of wetlands (see Giblett 1996)). Yosemite National Park was a paradise for Muir (1991, p. 76) 'that makes even the loss of Eden seem insignificant'. The Park contained parks which were as 'fair as Eden' (Muir 1991, p. 129), though perhaps Yosemite National Park was more what Roderick Neumann (1995, p. 155) has called in relation to national parks in Africa an 'ersatz Eden'.

Yet like the cinema with its modern technological simulation of the cathedral's stained-glass windows in which, as Virilio (1989, p. 38) puts it, 'everything visible appears to us in the light', 'the standard perception of national parks' was, as Alfred Runte (1979, p. 64) describes it, 'as a unique visual experience' centred on the aesthetically legitimated sense of sight. Like Wordsworth, Muir (1991, p. 2) bemoans the fact that national parks were subjected to what he calls 'the scenery habit' common among his contemporaries. Like Wordsworth too, he was prone to what could be called the sublimatory habit, especially in relation to *mountainous* national parks. The sublimatory habit transcended the common scenery habit and placed the poet and preservationist above his contemporaries as an arbiter of high cultural taste in landscapes.

Yet by functioning as reconsecrated modern cathedrals *forested* national parks could reverse history, or at least hark back to an earlier era, by returning the cathedral to the sacred, pre-Christian, pre-patriarchal grove from which it sprang. The cathedral's 'soaring columns and vaults', as Rupert Sheldrake (1991, pp. 44–45 and 176) points out, 'recall sacred groves, and vegetation bursts out everywhere [. . . M]any churches and cathedrals in Europe were built on pre-Christian sacred sites'. The forested national park as cathedral not only re-appropriates the appropriated sacred grove but also could return the sacred place to its pre-architectural, pre-Christian wild(er) ness manifestation, to what Devereux Butcher (1969, p. 1) calls 'temples not built with hands' (though national parks could contain areas made and worked by indigenous hands), unlike 'the great cinema temples' as Virilio (1989, p. 31) calls them.

Yet like the cinema temples, Muir (cited by Cohen 1984, p. 192) found that 'the money-changers were in the temple' of the Yosemite. The forested and mountainous national park as temple needed its modern-day Jesus who would drive the moneychangers, the cash-spinners, out. One such Jesus was John Muir (1991, p. 209; see also p. 225) who did not need to be reminded that 'great trees and groves used to be venerated as sacred monuments and halls of council and worship' and 'the hills and groves were God's first temples' (Muir, 1987, p. 146). Muir regarded Sequoia forests in particular as 'majestic living temples, the grandest of Gothic cathedrals' (cited by Cohen, 1984, p. 195).

The forested national park as cathedral and temple has meant that, as Sheldrake (1991, p. 177) goes on to suggest, 'a religious experience of the wilderness has endowed many of the American national parks with a transcendental quality. For many who visit them, they are more than recreational areas; they are natural temples or sanctuaries.' John Muir has been ordained as prophet and evangelist of this new religion of naturalistic Campbellism, or simply sanctuarism: a new St John, writer of a new gospel, a new good news about the saving graces of wild(er)ness sanctuary; a new John the Baptist, crying out in and for the wild(er)ness preparing the way for Aldo Leopold and his saving gospel of a land ethic and a conservation aesthetic; a new St John the Divine proclaiming the apocalypse of a blighted and benighted urban 'civilization' bereft of wild(er)ness; and a new John Calvin converting God/Nature dualism into the craggy patriarchal sublime and predestined glaciers of what Muir (1991, p. 173) called 'the stern wilderness' of the mountains, and of the mountains alone. Muir even saw himself as John the Baptist: 'Heaven knows that John the Baptist was not more eager to get all his fellow sinners into the Jordan than I to baptize mine in the beauty of God's mountains' (cited by Cohen 1984, p. 259; see also p. 254). Worster (1993, pp. 194–195) also argues that Muir 'became a kind of frontier evangelist' and 'invented a new kind of frontier religion'.

There is some support for these views of Muir in his own writings on national parks. For Muir (1991, p. 31), the cathedral of the national park with forests or geysers can produce a contemplative mood in 'the calmest, stillest scenery' of its forests, or can render the onlooker 'awe-stricken and silent, in devout, worshipping wonder' in the face of the 'awful uproar' and 'tremendous outburst' of its terrifying, sublime, phallicized geysers (Muir 1991, pp. 40–41). For Muir (1979, p. 350), the forested national park is a public place of community worship, not a private cloister for individual meditation, or more precisely it is a place for 'both solitude and society'. Writing of Yosemite National Park, Muir (1991, p. 59) described how 'nearly all the park is a profound solitude. Yet it is full of charming company, full of God's thoughts, a place of peace and safety amid the most exalted grandeur and eager enthusiastic action, a new song, a place of beginnings abounding in first lessons on life'. The monumental and the mountainous again appropriate the beginnings of life.

Forested national parks are also sacred places for Muir because they are a refuge or sanctuary, like the medieval monastery, for the fugitive. For Muir (1979, p. 350), Yosemite National Park is 'a place of rest, a refuge from the roar and dust and weary, nervous, wasting work of the lowlands', unlike the uplifting, sublimating highlands (Muir was, of course, of Scottish descent and in fact was born in Scotland). In an increasingly urbanized world where it seems the possible only other refuge from 'rust and disease' (Muir 1991, p. 1) is the golf course, but whose membership fees are so prohibitively expensive for most (the golf course is the rich man's private refuge), the forested national park is for Muir (1979, p. 352) 'the poor man's [public] refuge', but not for him the indigenes' working land. Muir was very much a proponent of the sanctuary idea for national parks that would provide a refuge for native flora and fauna, and for the poor lowlanders, or city dwellers, but not a site of inhabitation and enjoyment of ownership and rights by indigenous peoples.

Perhaps the poorest 'man' culturally and spiritually of all in the modern and hypermodern worlds is the tourist, or at least more precisely the modern mass-packaged touristic consumer. For Muir (1991, p. 9), 'the wildest health and pleasure ground accessible and available to tourists seeking escape from care and dust and early death are the parks and reservations of the West'. National parks can provide the tourist with an escape route from the prison of modern city life provided once they gets there they slow down. Muir (1991, p. 42) rails that 'nothing can be done well at a speed of forty miles *a day* [my emphasis]. The multitude of mixed, novel impressions rapidly piled on one another make only a dreamy, bewildering, swirling blur, most of which is unrememberable'. If the tourist treats the national park like the city with its flood of impressions, the result will be the same, undistinguished, unremarkable blur of experience.

The tourist also needs to experience the national park alone. For Muir (1991, p. 210), 'little [. . .] is to be learned in confused, hurried tourist trips, spending only a poor noisy hour in a branded grove with a guide. You should go looking and listening alone on long walks through the wild forests and groves in all the seasons of the year'. The sublime is a solitary experience. A national park is a cathedral because it takes time to appreciate its vastness, its fastnesses, its nooks and crannies. In Yosemite National Park, for Muir (1991, p. 58), 'nowhere will you see the majestic operations of nature more clearly revealed beside the frailest, most gentle and peaceful things'. For Muir (1991, p. 30) Yosemite National Park is not only a means to escape from an early death, but also 'the scenery is wild enough to awaken the dead'. The scenery can even shake the apathetic out of their lethargy. The geysers and hot springs of Yellowstone National Park display 'an exuberance of color and strange motion and energy calculated to surprise and frighten, charm and shake up the least sensitive out of apathy into newness of life' (Muir 1991, p. 31). Geysers for Muir are terrifying, sublime agents.

Geysers are also masculinized by Muir. The sublime in the masculinist modern Western tradition has been associated inextricably with the masculine virtue of what Terry Eagleton (1990, p. 54) calls 'virile strenuousness'. The sublime is, as Eagleton goes on to argue, 'on the side of enterprise, rivalry and individuation' and is 'the lawless masculine force which violates yet perpetually renews the feminine enclosure of beauty'. For Muir (1991, p. 214; see also p. 228), for example, the 'Big Tree' of the Sequoia shows 'Nature's immortal virility'. Similarly, geysers for Muir (1991, p. 31) are 'like inverted waterfalls', but also like the upright mountain in that they are phallic as Muir (1991, pp. 31–32) shortly leaves his reader in no doubt as he describes the geysers of Yellowstone National Park as 'standing rigid and erect, hissing, throbbing, booming'.

Muir (1991, p. 33) generally prefers geysers to waterfalls, both to swamps, and mountains over all three. For Muir (1991, p. 4) geysers are 'the laboratories and kitchens, in which, amid a thousand retorts and pots, we may see Nature at work as chemist or cook, cunningly compounding an infinite variety of mineral messes'. Nature is here figured ambiguously, even androgynously, as chemist or cook, even alchemist or witch, working in the laboratory and kitchen with thermodynamic technology in 'the hot underworld' as God works in His glacial workshop, the industrial inferno to produce meals (me(t)als?) for modernity. Nature as figured in these terms is not all that far away from the culture of what Muir (1991, p. 2) calls 'these hot, dim, strenuous times', these modern times in which base matter is sublimated into ethereal commodities using heat (though they are also 'these cold, doubting, questioning, scientific times' (Muir 1991, p. 41) produced by the penetrating light of reason.

God's national parks

Besides a view of nature sublimated in national parks, Muir also propounded a view of God-the-Father-nature incarnated in mountains, or mountainous scenery, with its dawn lights and alpenglows as a kind of God-the-Son figure. For Muir (1991, p. 56) 'next to the light of the dawn on the high mountain tops, the alpenglow is the most impressive of all the manifestations of God'. The alpenglow, sublimated above the mountains themselves, could manifest God, and mountains could even be God. When Muir (1980, pp. 18–19) encounters the Fairweather Mountains of Alaska he regards them as God Himself: 'here the mountains themselves were made divine'. For Muir, the mountains *were* God.

If the mountains could be God, and God could be incarnated in the alpenglow, the mountaineer could incarnate the mountains, as when Muir (1987, p. 15) states 'we are now in the mountains and they are in us' to the point of merging with nature in a romantic, oceanic feeling in which Muir (1980, p. 88; see also 1992, pp. 212) could claim that 'you lose consciousness of your own separate existence: you blend with the landscape, and become part and parcel of nature'. Even visitors to mountainous national parks could partake of a sacramental repetition or re-enactment of this incarnation. Muir (1991, p. xvii; my emphases) announced in his 'Preface' to his book on national parks that its aim was 'to show forth the beauty, grandeur, and all-embracing usefulness of our wild *mountain forest* reservations and parks with a view to inciting people to come and enjoy them, and get them into their heart, that so at length their preservation and right use might be made sure'. Muir makes no attempt to disguise the fact that his book is devoted to 'our' *mountain forest* national parks.

While subscribing to some dominant discourses of national parks, Muir also flies in the face of others, such as the national parks as useless or worthless lands (the so-called 'worthless lands' thesis (see Runte 1979, pp. 28, 49 and 60)), and national parks as nationalist parks in which America saw and commemorated itself as what Perry Miller (1956, p. 209 and 1967, p. 201) calls 'nature's nation'. Against both discourses, Muir advocates a kind of spiritual and sacramental incarnation of nature into the national heart or psyche unlike the material incorporation of nature into and by the gaping maw of the national, capitalist state and industrial technology (see MacCannell, 1992, p. 115). Yet as nature's nation also saw itself as God's nation, national parks were not only God's national parks for Muir but also for the nation at large. Yet the view of national parks in the United States as 'America's secular cathedrals' persists with a previous Secretary of the Interior, Bruce Babbitt (cited in Mitchell 1994, p. 12), describing them exactly in these terms.

Not only has 'Nature become a secular deity in this post-romantic age' as William Cronon (1996, p. 36) puts it, but national parks are America's secular cathedrals in which this deity is worshipped under the sign of the sublime. Muir's work in particular is caught up, as are national parks in general, in this triangular force field between, even in this American cultural and political triumvirate of, God, nature and nation with the national park as their point of intersection in what could be called 'God's nation's nature'. The American national park is thus not nature in the raw but nature cooked by culture. Muir repeats what Lawrence Buell (1995, p. 62) calls 'the gesture of putting the nation under the sign of the natural: America as crag, Canada as iceberg, Australia as outback'.

For Muir, national parks as nature incarnated, rather than nature incorporated, will not merely be places out there to 'enjoy' but places to which 'people' (though the question is raised of *which* people? as is the question of *whose* national parks? in relation to the title of his book – who does 'our' refer to?) belong in a *natural* nationalism and not so much places which belong to the people in *cultural* nationalism. By taking this position Muir sought to reverse the human/nature dualism so endemic to Western culture, even subvert the God/nature (including humans) dualism of Calvinism, and replace the democratic ideal with a nature-cratic one ruled by 'the parliament of trees' and the law of Nature, rather than the law of God, and supersede cultural nationalism with natural internationalism which would transcend nations and national borders.

Muir also sought to overcome a narrowly utilitarian view of forest reservations, the predecessor of national parks. Against the prevailing view of his time that mountainous landscapes were 'worthless lands' and that forest reservations were only useful for timber, Muir (cited by Runte 1979, p. 62) posed what he called 'the use of beauty'. Of a mountain wild(er)ness experience in general he avowed that:

> the tendency nowadays to wander in wilderness is delightful to see. Thousands of tired, nerve-shaken, over-civilized people are beginning to find out that going to mountains is going home; that wildness is a necessity; and that mountain parks and reservations are useful not only as fountains of timber and irrigating rivers, but as fountains of life.
>
> (Muir, 1991, p. 1; see also p. 217)

Muir equates wilderness with, and reduces it to, mountains. Muir also sees the mountains (and wilderness unlike his settler compatriots) as home.

Like one of the other 'grandfathers' of the American conservation movement in Henry David Thoreau (Tilden, 1970, p. 303 called Muir 'the Western Thoreau'), Muir praised the prophylactic features of wild(er)ness and a wild(er)ness experience. Muir followed the motto of the wilderness wanderer penned by Thoreau (cited by Hoagland, 1987, p. 48): 'in wildness is the preservation of the world'. Yet Thoreau viewed wilderness more broadly than Muir (1991, pp. 61, 70 and 74), who generally conflates wilderness, mountains and the sublime with their peaks figured as 'colossal spires' of 'massive sublimity'. This conflation is summed up in Muir's (1991, p. 70) reference to 'a sublime wilderness of mountains' and to 'the massy sublimity of the mountains' (Muir 1987, p. 141). Muir (1980, p. 124) also refers to 'a sublime wilderness of mountains' in his essays. By doing so Muir enacts a masculinist construction of nature and reality in which the glacial, masculine, homely and sublime of the mountains are seen as the origin of life rather than the littoral, feminine, matrifocal (un)homely and slime of the swamp or other wetland.[7]

Yellowstone National Park impressed Muir so much that he regarded climbing its mountains as a sublime experience. Climbing Electric Peak for Muir (1991, p. 44) is a means to 'get yourself kindly shaken and shocked. You are sure to be lost in wonder and praise, and every hair of your head will stand up and hum and sing like an enthusiastic congregation.' Muir, like Walt Whitman, sings the body electric (or more precisely, it sings him) of the individual, masculinized human body in a patriarchal hymn to God the Father Law, not to 'Mother Nature'. The sublime is an individual, not a communal, experience.

For Thoreau and Muir wild(er)ness, rather than the primitive or the barbarian with their pejorative overtones, was distinct from civilization (Muir, 1991, p. 2). In between the wild(er)ness of mountainous national parks and the civilization of the cities Muir (1991, p. 2) places 'the half wild parks and gardens of towns', though for him national parks are 'worth infinitely more than all the gardens and parks of town' (Muir, 1979, p. 351). Muir (1991, p. 2), however, sees this care for the parks and gardens of town as encouraging. He even finds encouraging 'the scenery habit in its most artificial forms, mixed with spectacles, silliness, and kodaks'. Yet the scenery habit is not only mixed with spectacles, silliness and kodaks but also potentially with the sublime and sublimation as both valorize the sense of sight over all other senses. It is perhaps hardly surprising then that Muir would not disparage the scenery habit but would hope it would lead on in similarly sighted fashion to the sublimatory habit.

If mountainous national parks are a mediating category for Muir (1991, p. 4) so too are human beings who he describes as 'half animal, half angel', half slime, half sublime, or s(ub)lime in Zoë Sofoulis' (1988, p. 12) parenthetical portmanteau. The angelic is strongly associated for Muir (1991, p. 56) with mountainous wild(er)ness that is 'the dwelling-place of the angels' whereas animals are 'fellow beings, so seldom regarded in civilization'; plants also for that matter will 'soon come to be regarded as brothers' (Muir, 1979, p. 350). Animals are even regarded by him as 'brimful of humanity' (Muir, 1991, p. 59) as is 'the whole wilderness [which] seems alive and familiar, full of humanity' (Muir, 1987, p. 238). Muir's assessment of plants, animals and wilderness smacks of anthropocentrism, but it may have been a tactical move to see them at least in human terms and not less than, or alien to, the human.

Or the Great Goddesses'?

Yet Muir also propounded a competing view of nature sacralized to that of his naturalistic Campbellism. He concedes that it is not just mountains and forests that are enjoyable and inspirational, but all wilderness. Muir stated categorically that 'none of Nature's landscapes are ugly so long as they are wild' (Muir, 1991, p. 4). As wildness is a matter of point of view, rather than a quality of substance (see Giblett, 2011, chapter 5), Muir would seem to subscribe to Thoreau's (1962, *VIII, p.* 275) dictum that 'the same object is ugly or beautiful, according to the angle from which you view it'. Indeed, according to the distance from, and the angle at which, we view it as (see Giblett, 2011, chapter 3) and our predisposition towards it as Thoreau (1962, *IX*, p. 466) maintained later that 'we find only the world we look for'. The world we look for is the world we find. We cannot find what we do not see.

Beautiful (because wild) landscapes could include swamps as Muir (1991, p. 5) discovered in his conversion to conservationism, or at least to sanctuarism: '*Calypso borealis* still hides in the arbor vitae swamps of Canada, and away to the southward there a few unspoiled swamps, big ones, where miasma, snakes, alligators, like guardian angels, defend their treasures and keep them as pure as paradise.' Yet rather than separating Muir off as an original figure, this conversion experience in the wilderness makes him fit the pattern of the frontier hero as outlined by Richard Slotkin (1994, p. 374). According to Wolfe (1945, pp. 146–147), Muir's encounter with the *Calypso borealis* in a Canadian swamp was one of 'the two supreme moments of his life', the other being his meeting with Emerson, and not his meetings with Roosevelt, the Sierras or sequoias. The swamps referred to as 'away to the southward' were presumably the swamps of Florida, Muir's encounter with which he devotes one chapter of his *A Thousand-Mile Walk to the Gulf* (as discussed previously in chapter 4 above; see also Worster, 2008, pp. 93–95).

Muir propounded not only a sublime aesthetics of the mountains, geysers and glaciers, but also a watery and earthy (slimy) experience of the wetlands. Muir (cited by Runte 1979, p. 31[8]) looked forward to the day when 'lowlands will be loved more than alps, and lakes and level rivers more than waterfalls'. That day has far from arrived. Mountains and waterfalls are still a dominant feature of the modern aesthetics of landscape. Yet lowlands for Muir (1987, p. 186) not only included wetlands, but also cities with their 'lowland care and dust and din, where Nature is covered and her voice smothered'. Perhaps he was implying that cities should also be loved more than alps and waterfalls.

Muir's opposition to the narrowly utilitarian point of view of nature and national parks can also be seen in his answer to the rhetorical question 'what are rattlesnakes good for?' Muir (1991, p. 25) replies that they are 'good for themselves' (as a national park should be looked after 'for its own sake' and 'we need not begrudge them [the rattlesnakes] their share of life' (Muir 1991, p. 43). Muir's position on rattlesnakes (and alligators) has been seen as the foundation for a biocentric ethic.[9] Muir's anthropocentrism in his aims for his book and in his attitude to plants and animals in general and his biocentrism in relation to rattlesnakes in particular were contradictory positions, but could also be seen as tactical moves depending on the object under discussion and the entrenched attitudes to them either to be extended (in the case of humanism to plants and animals) or reversed (in the case of rattlesnakes).

Yet by converting God into nature and by subscribing to biocentrism rather than to anthropocentrism, Muir was staying within and reproducing their categories. Mountainous, sublimated nature in national parks was still God the Father Law for Muir though in his pantheon he could also allow for a slimy, desublimated unnamed Palaeolithic Great Goddess in the swamps and a good Neolithic 'Mother Earth' or 'Mother Nature' everywhere. For Muir (1980, p. 229) 'Nature is a good mother' as well as a bad father, or more precisely nature for him is split between the two.

Muir's work on, or more precisely for, mountainous and/or forested, national parks marks a shift from an anthropocentric to a biocentric ethic for national park and wild(er)ness areas, though Muir's article on 'The American Forests' for *The Atlantic Monthly* (republished in *Our National Parks*) showed, as Roderick Nash (1982, p. 136) points out, 'Muir's continued ambivalence on the 'forestry-or-preservation issue''. Muir's other articles of the 1890s for *The Atlantic Monthly*, particularly 'The Wild Parks and Forest Reservations of the West' (also republished in *Our National Parks* (see Muir, 1991)), are far less ambivalent on this issue and far more committed to the preservation side of the struggle.

Muir saw that the wild(er)ness, and not just mountain wilderness, was already safer than civilization. The implication is that so-called 'civilization', not wild(er) ness, needs to be made safe. Despite the snakes and alligators and other natural predators of the swamp wild(er)ness, Muir (1991, p. 21) thought that 'it is far safer to wander in God's woods than to travel on black highways or to stay at home [. . .] No American wilderness that I know of is so dangerous as a city home "with all the modern improvements". One should go to the woods for safety, if for nothing else.' Perhaps one should not go to swamp for safety, not to the woods and mountains, though the latter could provide the terrifying and uplifting experience of the sublime, whereas the former could mean a horrifying and down-putting immersion in the slimy.

For Muir, one should go to the woods not only for physical safety but also for spiritual succour. In fact, for him 'what we call parks' would better be called 'places for rest, inspiration, and prayers' (Muir, 1991, p. 23). Wooded national parks for Muir are sacred places in which, as Rupert Sheldrake (1991, p. 23) suggests, 'the spiritual and the physical are experienced together. Sacred places are openings between the heavens and the earth.' Muir certainly saw mountainous and wooded national parks as sacred places in which the spiritual and the physical could be experienced together via a sublime opening between the heavens and the earth. He even saw national parks as sacred places where it would be possible to overcome the dualism between the physical and the spiritual. Indeed, on the one hand, he is largely a Platonist in his reference to 'the truly substantial, spiritual world whose forms flesh and wood, rock and water, air and sunshine, only veil and conceal' and, on the other, a naturalistic Campbellite in his preference for the immanence, or even incarnation, of the spiritual in the physical for 'here [in Yellowstone National Park] is heaven and the dwelling-place of angels. . . the terrestrial manifestations of God' (Muir 1991, p. 55–56).

Muir seems to have sensed no contradiction between these two positions expressed together within a page of each other because for him there was no contradiction between them. Muir saw an exchange and mutuality between the spiritual and the material exemplified in the forested and/or mountainous national park. He refused the views of the capitalists, scientists, engineers and technocrats of his time who saw nature as so much raw material and dead matter to be manipulated and exploited, and of the Christians who saw nature as so much clinging and sinful materiality to be sloughed off and transcended. For him nature was living, divine, dynamic yet it also had its stern and craggy, patriarchal and Campbellite God up in the mountains and its wise, tender, though also stern (see Muir, 1987, p. 142) and matrifocal Great Goddess down in the swamps. Nature for Muir was bifurcated, even trifurcated with the third figure of a good Mother Earth or Mother Nature.

Muir subscribed not just to 'the immanence of nature in the divine' as Sheldrake (1991, p. 197) puts it, but also to the transcendence of the divine in nature. Muir was thus, in Sheldrake's (1991, p. 198) terms, a *panentheist* for whom:

> God is not remote and separate from nature, but immanent in it. Yet at the same time, God is the unity that transcends it. In other words, God is not just immanent in nature, as in pantheist philosophies, and not just transcendent, as in deist philosophies, but both immanent and transcendent, a philosophy known as *panentheism*.

Or should that be Goddess, rather than God? Zimmerman (1994, p. 254) maintains that 'goddess spirituality holds that the divine is both immanent [. . .] and transcendent'. The divine for Muir overarches nature, transcends everything, but is immanent in individual, particular places, and not just in forested and/or mountainous national parks but also in swamps; the divine is cosmically and locally transcendent and immanent; the whole cosmos and every locality manifests the divine, the divine transcends the local, the divine is in the local, and vice versa. Muir was no mere radical who 'thought globally and acted locally', but a panentheist who thought the divine in the local and enacted the local in the divine.

Writing the word of nature

If national parks function as modern cathedrals with their 'sermons in stones' (Muir, 1991, p. 59), then their sacred scriptures are the inscriptions of nature *in* the earth (unlike the writing of the modern, colonial city *on* the earth and unlike the slimy tracery of the Great Goddess, the swamps, the womb of the earth (see Giblett 1996, especially chapter 4). Muir (1991, p. 45) advises his readers that 'after this reviving experience [of climbing Electric Peak], you should take a look into a few of the tertiary volumes of the grand geological library of the park [Yellowstone National Park], and see how God writes history'. Muir was the culmination of a nineteenth-century American phenomenon in which, as Perry Miller

(1956, p. 211 and 1967, p. 203) puts it, 'nature somehow. . . had effectually taken the place of the Bible'. Or more precisely, as nature had taken the place of God for Muir, so the Word of Nature had taken the place of the Word of God. Muir was a self-mythologized literary frontier hero who was, in Richard Slotkin's (1994, p. 374) terms, 'generally disinclined to learn from book culture when the book of nature is free to read before him'. Printing made the word of God available to the vernacular reader whereas Muir found the book of nature already open ready for him to read freely.

In the modern mountainous wild(er)ness one can see, not how God writes the law in words on tablets of stone as 'He' did in the Judaic wilderness with Moses, but how 'God' writes history in 'words' of stone in 'the divine manuscript' (Muir, 1987, p. 132) of mountainous nature as 'He' does in the law of Muir, the modern-day Moses of mountainous national parks. For Muir (1991, p. 45), here in Yellowstone National Park are 'a wonderful set of volumes lying on their sides, – books a million years old, well bound, miles in size, with full-page illustrations. . . telling wonderful tales of bygone centuries'. For Muir (1991, p. 59), the 'natural history', or better national park, or even better still mountainous wild(er)ness writer, 'to get all this ['the wonderful clearness and freshness of the rocky pages'] into words is a hopeless task'. As a result, 'to defrauded town toilers, parks in magazine articles are like pictures of bread to the hungry. I can write only hints to incite good wanderers to come to the feast' (Muir, 1991, p. 59). Rather than the oral sadistic feast of capitalism, the feast Muir had in mind here was more a Passover feast, a sacramental communion. Muir as a magazine article writer was the frontier hero of a literary mythology through whom, as Slotkin (1994, p. 374) puts it, 'the deep wisdom of the natural wilderness is transmitted to the sluggish life of the Metropolis'.

One course of the feast, to switch metaphors, would be a richly layered text. In these books of nature 'the post-glacial agents [of air, rain, frost, rivers, earthquakes, avalanches (Muir, 1991, p. 63)] are at work on the grand old palimpsest of the park region, inscribing new characters' (Muir 1991, p. 49) on the *tabula rasa* or 'white bosom' of the glaciers rather than tracing stories in the living black waters of wetlands. The book of nature is not set in stone for all time but is constantly changing, with new writings being inscribed over old ones to produce a deeply and richly patterned palimpsest. Muir's own writing on glaciers is one layer of the palimpsest inscribed on its broad white bosom. Glaciers themselves may deserve 'the name of Destroyer. . . but we quickly learn that destruction is creation' (Muir, 1991, p. 72) and 'we see that everything in Nature called destruction must be creation (Muir, 1987, p. 229)'. That natural (not human) destruction is creation is one of Muir's favourite maxims (see Muir, 1987, p. 238 and 1980, pp. 31and 53).

Not only glaciers and post-glacial agents write the text of the book of nature, but all inhabitants of the wild(er)ness do, even the swarming insects are 'a cloud of witnesses telling Nature's joy' and the plants are 'singing the old new song of creation'. All creatures have a kind of evangelistic function of spreading the gospel according to John Muir (1991, pp. 52–53) about the religion of nature. Indeed, 'each and all tell the orderly love-beats of Nature's heart', 'the heart of nature, whence we came' (Muir, 1980, p. 46), couched within 'the generous bosom of the woods' (Muir, 1992, pp. 155–156), nature's good breast rather than nature's bad breast of the black waters of wetlands.

Similarly, for Muir (1991, p. 57) in terms reminiscent of Gerard Manley Hopkins, the Sierra Nevada mountains in Yosemite National Park are 'pervaded with divine light, every landscape glows like a countenance hallowed in eternal repose; and every one of its creatures [. . .] is throbbing and pulsing with the heartbeats of God'. This creative work of nature is always going on with destructive work: 'Nature is ever at work building and pulling down, creating and destroying, keeping everything whirling and flowing, allowing no rest but in rhythmical motion, chasing everything in endless song out of one beautiful form into another' (Muir 1991, p. 73). Nature here is always at work producing not so much a cosmos as a cosmogenesis. Muir's work here is remarkably prescient, even prophetic, of the paradigm shift from modern nature as (dead) machine to postmodern nature as living community, even communion (see, for example, Swimme and Berry, 1992).

Nature is a homely living agent for Muir rather than unhomely dead matter or (un)homely Great Goddess. Going to the woods for Muir (1991, p. 74) is going home, or 'going to the mountains is like going home' (Muir, 1980, p. 118); going to a national park (but not going to a swamp) for him is 'getting in touch with the nerves of Mother Earth' (Muir, 1991, p. 2). Nerves communicate with the whole body through electrical impulses, so getting in touch with the nerves of the body electric of Mother Earth meant getting in touch with the fibres that communicate with the whole body of Mother Earth, both inside and outside the national park.

Yet getting in touch with the nerves of Mother Earth meant *not* getting in touch with her whole body. Going *into* the swamp would be getting in touch with the body of Mother Earth. Perhaps it was in going to the mountains, rather than to the national park per se, that Muir got in touch with the sublimated body of Father Law. The mountains could be incarnated in Muir, and Muir in the mountains, so that there was no disconnection between the two and he became 'an inseparable part of it [. . .], a part of all nature' (Muir, 1987, p. 16). In 'the Sierra Cathedral', for Muir (1987, p. 243), 'we feel ourselves part of wild Nature, kin to everything'.

If Mother Earth for Muir has nerves she also has a heart, what Muir (1991, p. 4) calls 'the warm, unspoilable heart of the earth', 'Nature's warm heart' (Muir, 1991, p. 180). For Muir, the warm, good breast of 'Mother Earth is ever familiar and the same' (Muir, 1991, p. 37), unlike the hot workshops of geysers where Father God Law is ever unfamiliar and different, and unlike the hot netherlands of swamps where the Great Goddess is ever uncanny and (un)homely. 'Nature's sources never fail' (Muir, p. 1991, p. 42). The good breast of what Muir calls in a letter the 'bosom' of 'sweet kindly Mother Earth' (Badè, 1924, *II*, p. 267) never dries up and the (glacial and/or wetland?) womb generously gives birth demonstrating 'the infinite lavishness and fertility of Nature – inexhaustible abundance' (Muir, 1987, p. 242). Nature, for Muir (1980, p. 229), and in short, is 'a good mother', though also a bad Father (like his own about whom we hear so much in the biographies while Muir's 'Mother Earth' is perhaps like his mother about whom his readers hear so little, except for his surrogate mother in Jeanne Carr). This lack of 'a parental tie to anchor him to the Metropolis', or to metropolitan culture more generally, helps to make Muir fit Slotkin's (1994, p. 374) model of the frontier hero. Muir (1991, p. 268) prefers the warm, living nature of Mother Earth, a nurse maid nature 'always ready to heal every scar' 'one touch' of whose hand 'makes the whole world kin' (Muir, 1991, p. 164), to 'the gray, savage wilderness of crags and peaks', the visage of his mountainous God the Father Law, who 'seems lifeless and bare' (Muir, 1991, p. 125).

If Mother Earth has nerves and a heart, then Father Mountain has a face or more precisely a 'divine landscape-countenance' with 'every feature glowing' (Muir, 1987, p. 15 and 115). For Muir, Father God mountain is the body of nature from the neck up, the site of sublimation, intellection and theorization, the heights of the sublime spires, whereas the Great Goddess and Mother Earth are the body of nature from the neck down, the womb of new life, the organs of nurturing, the depths of the slimy swamps. Muir (1991, p. 55) remarks on how 'the physiognomy and even the complexion of the landscape should still be so divinely fine! and 'all the landscape is glowing like a benevolent countenance at rest' (Muir, 1991, p. 193) or 'the landscape [is] beaming with consciousness like the face of a god' (Muir, 1987, p. 85), when referring to the sublimated alpen glow. Muir looks on this face in mountainous and forested national park, the stern, craggy face of God the Father Law and not on the body of the Great Goddess, life-giving and death-dealing.

Notes

1 See Worster (1993, pp. 190–196) for a useful discussion of Muir's more specifically Campbellite upbringing. Worster bemoans the fact that Fox (1981) and Cohen (1984) do not even mention Campbellism and that Turner (1985, pp. 192–193) only treats it in passing. See also Worster (2008, pp. 37–39).
2 According to Worster (1993, p. 240, n20), the Calvinist doctrine of predestination would have been elitist for the Campbellite Daniel Muir, John's father, though John saw glaciers as predestined. See below.
3 'Mother Earth' for Muir is not strictly speaking the Neolithic earth mother of agriculture and patriarchy, nor is it explicitly the Palaeolithic swamp Great Goddess of 'matriarchy'.
4 I discuss cinemas as cathedrals of industrial capitalism in Giblett (2008, chapter 5).
5 For mountains as monuments for Muir see Muir (1980, p. 35) and for monumentalism as the impetus behind the national park idea see Runte (1979, pp. 28, 49 and 60).
6 For Yosemite as sublime see Muir (1991, p. 67 and 70; 1987, especially chapter 5, 'The Yosemite', where he also repeatedly refers to the Yosemite temple; and 1980, p. 97). Other forests could also be sublime cathedrals, as in Alaska where Muir (1980, pp. 48–49) was 'charmed with the majestic beauty and grandeur of the trees, as well as with the solemn stillness and the beauty of the elastic carpet of golden mosses flecked and barred with the sun-beams that sift through the leafy ceiling'.
7 I discuss the s(ub)lime and the wetland as womb of the world in Giblett (1996).
8 I discuss Muir's view of wetlands and swamps in Giblett (1996, chapter 10).
9 I discuss Muir's encounter with an alligator in chapter 4 above.

References

Badè, W. 1924. *The Life and Letters of John Muir, Volumes I and II*. Boston: Houghton Mifflin.

Benjamin, W. 1986. *Moscow Diary*. G. Smith, ed., R. Sieburth, trans. Cambridge, MA: Harvard University Press.

Bookchin, M. 1991. *The Ecology of Freedom: The Emergence and Dissolution of Hierarchy*. Revised Edition. Montreal: Black Rose.

Buell, L. 1995. *The Environmental Imagination: Thoreau, Nature Writing, and the Formation of American Culture.*, Cambridge, MA: Harvard University Press.

Butcher, D. 1969. *Exploring Our National Parks and Monuments*. 6th Edition, Revised. Boston: Houghton Mifflin.

Cohen, M. 1984. *The Pathless Way: John Muir and American Wilderness*. Madison: University of Wisconsin Press.

Cronon, W. 1996. Introduction: In Search of Nature. *In*: W. Cronon, ed., *Uncommon Ground: Rethinking the Human Place in Nature*, New York: W. W. Norton, pp. 23–66.

Day, D. 1996. *Claiming a Continent: A History of Australia*. Pymble, NSW: Angus and Robertson.

Eagleton, T. 1990. *The Ideology of the Aesthetic*. Oxford: Basil Blackwell.

Flichy, P. 1995. *The Dynamics of Modern Communication: The Shaping and Impact of New Communication Technologies*. L. Libbrecht, trans. London: Sage.

Fox, S. 1981. *John Muir and his Legacy: The American Conservation Movement*. Boston: Little, Brown.

Giblett, R. 1996. *Postmodern Wetlands: Culture, History, Ecology*. Edinburgh: Edinburgh University Press.

Giblett, R. 2008. *Sublime Communication Technologies*. Basingstoke: Palgrave Macmillan.

Giblett, R. 2011. *People and Places of Nature and Culture*. Bristol: Intellect Books.

Giblett, R. 2015. *Canadian Wetlands: Places and People*. Bristol: Intellect Books.

Gilpin, W. 1794. *Three Essays: on Picturesque Beauty; on Picturesque Travel; and on Sketching Landscape* [. . .] Second edition. London: A. Blamire.

Grove, R. 1995. *Green Imperialism: Colonial Expansion, Tropical Island Edens and the Origins of Environmentalism*. Cambridge: Cambridge University Press.

Hoagland, E. 1987. In Praise of John Muir. *In*: D. Halpern, ed., *On Nature: Nature, Landscape, and Natural History*, San Francisco: North Point Press, pp. 45–58.

Lane, B. 1988. *Landscapes of the Sacred: Geography and Narrative in American Spirituality*. Mahwah, NJ: Paulist Press.

MacCannell, D. 1992. Nature Incorporated. *In: Empty Meeting Grounds: The Tourist Papers*, London: Routledge, pp. 114–117.

MacKenzie, J. 1988. *The Empire of Nature: Hunting, Conservation and British Imperialism*. Manchester: Manchester University Press.

Miller, P. 1956. Nature and the National Ego. *In: Errand into the Wilderness*, Cambridge, MA: Belknap Press of Harvard University Press, pp. 204–216.

Miller, P. 1967. The Romantic Dilemma in American Nationalism and the Concept of Nature. *In: Nature's Nation*, Cambridge, MA: Belknap Press of Harvard University Press, pp. 197–207.

Mitchell, J. 1994. Our National Parks, *National Geographic*, 186(4), pp. 2–55.

Muir, J. 1979. Thoughts Upon National Parks. *In*: L. Wolfe, ed., *John of the Mountains: The Unpublished Journals of John Muir*, Madison: University of Wisconsin Press, pp. 350–354.

Muir, J. 1980. *Wilderness Essays*. F. Buske, ed. Salt Lake City: Peregrine Smith.
Muir, J. 1987. *My First Summer in the Sierras*. New York: Penguin. First published 1911.
Muir, J. 1991. *Our National Parks*. San Francisco: Sierra Club Books. First published 1901.
Muir, J. 1992. *A Thousand-Mile Walk to the Gulf*. New York: Penguin. First published 1916.
Nash, R. 1982. *Wilderness and the American Mind*. Third edition. New Haven: Yale University Press.
Neumann, R. 1995. Ways of Seeing Africa: Colonial Recasting of African Society and Landscape in Serengeti National Park, *Ecumene*, 2, pp. 149–169.
Runte, A. 1979. *National Parks: The American Experience*. Lincoln: University of Nebraska Press.
Schama, S. 1995. *Landscape and Memory*. London: HarperCollins.
Sears, J. 1989. *Sacred Places: American Tourist Attractions in the Nineteenth Century*. New York: Oxford University Press.
Sheldrake, R. 1991. *The Rebirth of Nature: The Greening of Science and God*. New York: Bantam.
Slotkin, R. 1994. *The Fatal Environment: The Myth of the Frontier in the Age of Industrialization 1800–1890*. New York: HarperCollins.
Sofoulis, Z. 1988. *Through the Lumen: Frankenstein and the Optics of Re-Origination*. Ph.D. Thesis, History of Consciousness, University of California, Santa Cruz.
Swimme, B. and Berry, T. 1992. *The Universe Story: From the Primordial Flaring Forth to the Ecozoic Era – A Celebration of the Unfolding of the Cosmos*. San Francisco: Harper.
Thomas, K. 1984. *Man and the Natural World: Changing Attitudes in England 1500–1800*. Harmondsworth: Penguin.
Thoreau, H. 1962. *The Journal 1837–1861, Volumes I–XIV*. B. Torrey and F. Allen, eds. New York: Dover.
Thoreau, H. 1980. *Natural History Essays*. Salt Lake City: Gibbs Smith.
Tilden, F. 1970. *The National Parks*. Revised and Enlarged Edition. New York: Alfred A. Knopf.
Turner, F. (1985). *Rediscovering America: John Muir in his Time and Ours*. New York: Viking.
Virilio, P. 1989. *War and Cinema: The Logistics of Perception*. P. Camiller, trans. London: Verso.
Wolfe, L. 1945. *Son of the Wilderness*. New York: Alfred A. Knopf.
Worster, D. 1993. *The Wealth of Nature: Environmental History and the Ecological Imagination*. New York: Oxford University Press.
Worster, D. 2008. *A Passion for Nature: The Life of John Muir*. Oxford: Oxford University Press.
Zimmerman, M. 1994. *Contesting Earth's Future: Radical Ecology and Postmodernity*. Berkeley: University of California Press.

8 Looking back on destruction and being at home in one's time, body and place

Lot's wife and the Angels of History, Geography and Corporeality

Unlike traditional human beings who have a cyclical sense of time, modern urban human beings are always performers and players in time suspended between a past we cannot change, a present we cannot escape from and a future we cannot know. We can only know the past. We can know something of the present on the cusp between the known past and the unknown and unknowable future. We can live in hope, a state of being and living, in the present for the future, learning from the past (but not dwelling nostalgically in it or looking fixatedly back at it), by facing and living in the present, and by having hope for the future (and not despairing about it). We can thereby seek not to be writers, readers and victims of the history of the past, but agents and writers of the history of the present (to use Michel Foucault's terms (1977, p. 31)).

Drawing on the biblical story of Lot's wife, who was turned into a pillar of salt for looking back at the destruction of the cities of Sodom and Gomorrah, and on Walter Benjamin's 'Angel of History' looking back at the wreckage of the past with her back turned the future and using his concept of the dialectical image between 'the what-has-been' and 'the fulfilled now' in a history of the present, this chapter develops the dialectical image between 'the what-has-been' and 'the fulfilled here' in a geography of places with what could be called the 'Angel of Geography' looking over wrecked and lost places. It also develops what could be called the 'Angel of Corporeality' who has and practices a non-exilic 'bodily presence of mind' (to also use Benjamin's terms) in place.

This chapter does so by: first, comparing and contrasting Benjamin's and his near contemporary Stefan Zweig's loss of their home places in European cities and their experience of exile from them; secondly, invoking Benjamin's and Henry David Thoreau's sense of being at home in one's body and place; and thirdly, considering the destruction of European cities literally or culturally as home places before or during World War II, often construed in terms of the destruction of Sodom and/or Gomorrah, but rarely invoking Lot's wife and the injunction not to look back, which this chapter begins by discussing.

Lot's wife

As the biblical book of *Genesis* relates, Lot's wife, in wilful disobedience of God's interdiction, looks back with nostalgia (regret and longing) at God's destruction of the sinful cities of Sodom and Gomorrah and so God turns her into a pillar of salt in capital punishment (as He, a being of His word, or more precisely, the being of the Word, said He would do) (*Genesis* 19: 17 and 26). This story can be read as a parable, moral fable and salutary warning that all who look back fixatedly into the past with longing and regret will be turned into a dead, dry and immobile pillar. It is a stark reminder of what *not* to do. 'The injunction not to look back', J. A. Loader (1990, p. 41) comments in his exegesis of the biblical story, 'is a widespread motif found often in the folklore of widely differing cultures'. As Loader's study shows, the Sodom and Gomorrah stories in *Genesis* 18–19 'inspired a rich tradition in early Jewish and Christian literature, but which also [. . .] shows [*sic*] signs of the existence of earlier Sodom and Gomorrah traditions' (Loader, 1990, p. 13; see also pp. 139–140). The stories and the injunction not to look back also inspired a rich tradition in art, film and photography, as Martin Harries' (2007) recent study of all three media shows.

By contrast with Lot's wife who looks back (to the past, to a destroyed place, to a lost time and place) and is destroyed for doing so, the Angel of History for Walter Benjamin looks back into the past and the destruction that took place in it with her back turned towards the future and with the winds of progress beating into her face and pushing her into the future. Rather than 'a chain of events that appears before us', the Angel of History for Benjamin (2003, p. 392) 'sees one single catastrophe which keeps piling wreckage upon wreckage and hurls it at his feet'. The angel is 'driven irresistibly into the future to which his back is turned, while the pile of debris before him grows toward the sky'. Rather than a series of catastrophes proceeding chronologically one after the other through history, Benjamin's 'Angel of History' sees one single catastrophe of wreckage in time. This single catastrophe, however, occurs not only in time, in history, but also in space, in a place, in geography (as the biblical story shows). The 'Angel of Geography' sees one single catastrophe of wrecked and lost places, such as wetlands dredged, filled and so 'reclaimed', cities and suburbs set in them and being rereclaimed by them in storms and floods (and increasingly so in the age of global warming), and cities ruined in destructive cataclysms, such as Sodom and Gomorrah, but more in watery ones than fiery ones.

As well as the Angels of History and Geography – in one case whose divine and eternal point of view of wreckage is unattainable by mere mortals and in the other whose omniscient and omnipresent point of view of wrecked places is also unattainable by mere mortals – 'the Angel of Corporeality' lives bodily present in mind and place, a non-dualistic, holistic way of being probably unachievable by modern human beings. While the Angel of History turns 'the threatening future' into 'a fulfilled now' as Benjamin (1996, p. 483) puts it, the Angel of Geography turns threatening places into a fulfilled here and the Angel of Corporeality turns the threatening mind and/or body into a fulfilled, whole mind-body. From this point and place in space-time, these angels look over disastrous places and ruined bodies and minds, as well as look back at the disastrous past with mixed fear of, and nostalgia, for the past and lost places. They are not afraid and they are in no danger of being turned into a pillar of salt, like Lot's wife, for they are not fixated on the past. They face the past, present, places and the future with fear and hope in embodied minds and mindful bodies.

Looking back

Looking back, or not, into the past is a pressing issue for the exile from home, such as Benjamin and for his near contemporary Stefan Zweig, who were displaced not only from their homes and home cities, but also from the past (and from their own bodies and homeland). In his recent study of the life and work of Stefan Zweig entitled *The Impossible Exile*, George Prochnik (2014, p. 12) suggests that Zweig, 'the quintessential exile manqué, offers a formula for toxic migration – what might be called "Lot's wife syndrome"' as 'he could not stop looking back over his shoulder'. As Lot's wife was turned into a pillar of salt, Prochnik implies that Zweig in looking back over his shoulder experienced a similar fate such that he was locked into looking back into the past to the detriment of looking at the present and to the future, and was thereby atrophied or petrified (literally made into stone). A closer reading of the life and work of Zweig, however, shows a more nuanced story.

Zweig certainly recoiled in horror at the Nazi atrocities that he was living through in the present, as so many did then and since, but he also looked to a better future. Zweig may have suffered to some extent from 'Lot's wife syndrome' in his nostalgia for the artistic glories and luxurious comforts of 'Old Europe', but he was also Janus-faced, looking both back and forward, acknowledging the horrific realities of the present and espousing hope for the future of a 'New Europe'. Zweig's (2014) autobiography or memoir, *The World of Yesterday*, not only looks back to the past from the present, but is also 'a message to the future', as Prochnik (2014, p. 358) puts it later in the book. Zweig's (2016) recently translated collections of essays, *Messages from a Lost World: Europe on the Brink*, not only looks back to the history of yesterday, such as in 'The Vienna of Yesterday', but also forward to the history of tomorrow, such as in 'The Historiography of Tomorrow'. Zweig in exile was not only, as Volker Weidermann (2016, p. 145; my emphases) suggests, 'a *flâneur from* Old Europe on the edge of war', but also a *flâneur of* Old Europe in time and space, and a *flâneur for* New Europe.

Zweig was not only a *flâneur* who went strolling literally and figuratively through Old Europe and in hope for New Europe, but was also an aesthete who professed and practiced 'a fanaticism for art' as Prochnik (2014, p. 5) puts it, and thereby offers a formula for toxic aestheticism – the aestheticization of art and the practices of everyday life, and the sublimation of politics. Zweig believed in what Prochnik (2014, p. 343) later calls 'the capacity of art of induce to a sense of earthly transcendence – all woes and petty factionalism sublimated in aesthetic rapture'.

Zweig can be contrasted here with Benjamin, the quintessential nomadic exile and *flâneur* of culture (Old and New, European and not, in arcades and libraries, in cities and streets, etc.), and his stance on fascism and communism, and their contrasting fanaticisms or fascinations for art and politics. 'The logical outcome of fascism', for Benjamin (2003, pp. 269 and 270), is 'an aestheticizing of political life' to which 'communism replies by politicizing art'. A case in point here is Benjamin's communist or historical materialist reading of Kafka's desublimated and embodied art in Benjamin's terms of a dialectical image (to which I return below). Aestheticism aestheticizes art and everyday life. Art is cut off from everyday life and the body, and sublimated into the transcendent realm of the mind and the aesthetic.

In his biographical and critical study of the life and work of Zweig, Prochnik (2014, p. 147) cites Gershom Scholem, who traced the differences between Benjamin and Zweig and other German-speaking Jewish exiles in terms of their relationship to German people and language. In an essay on Benjamin, Scholem (2012, p. 19) argued that Zweig and some other Jewish writers were especially susceptible to what he called 'a lurid and tragic illusion' of belonging to the German people, whereas Benjamin, Freud and Kafka, even while being 'tied to the German language and its intellectual world', 'never succumbed to the illusion of being at home' because 'they were not of this world at all'.

By contrast, Zweig was a man of this world with all its perks and privileges, status and pleasures, made possible by his wealth. He was 'a man of the world' in a way that Benjamin and Kafka never were (and Thoreau certainly was not), and that Freud aspired to be and occasionally attained. Yet Prochnik's (2014, p. 354) later claim that Zweig 'never really felt at home anywhere', not only puts him at least partially in Scholem's other camp of Benjamin, Freud and Kafka, but also seems overdrawn as Zweig did feel at home in his house in the city of Salzburg before he left it because of the Nazis, and he had felt at home in Vienna in his early years, and in Paris after that. Exile made Zweig homeless and perhaps unable to feel at home anywhere else after that, whereas Benjamin, Freud and Kafka were never at home anywhere, or were equally at (un)home everywhere (and so were connoisseurs and proponents of the uncanny, or (un) homely in Freud's (2003) terms), including Benjamin during his childhood in Berlin and his exile in Paris and other parts (whereas Thoreau was always at home in his home place of Concord village and River, Walden Pond and Woods, various local swamps, and on the other nearby rivers with their confluence in Concord).

Not only had exile for Zweig according to Prochnik (2014, p. 254) 'forced him to live out an idea he'd once entertained as a philosophical tenet', but also, as Weidermann (2016, p. 37) puts it, 'for a man like him, a threatened state of exile was fatal', as it indeed proved to be. Zweig was forced to live out an idea of exile in and as a fatal state. In this respect, he is not that different from Benjamin. Both committed suicide in exile, a fatal state in a variety of senses. Both lived to some extent in exile from their own bodies (to which I return below). Both were stateless politically and corporeally. Benjamin looked back critically and nostalgically to traditional human beings who were at home in their own bodies and not exiled from them, like modern urban beings. The Angel of Corporeality performed the miracle of turning the body (portending disaster, decay and death) into the fulfilled 'now' of bodily presence of mind, present in time and space, in place.

Prochnik's statement that Zweig was always looking back over his shoulder like Lot's wife (and so was turned into a pillar of salt fixated on the past, denying the present and avoiding the future) seems overdrawn in relation especially to Zweig's relationship to exile, travel and home, and his accounts thereof. In his journal entry for 27th September 1935, Zweig (cited by Stone, 2011, p. vii) expressed the sentiment that 'one is everywhere at home'. He went on to praise 'the art of leaving behind oneself, without sentimentality, a good portion of one's past'. This belies to some extent Prochnik's pronouncement that Zweig 'could not stop looking back over his shoulder' and suffering from 'Lot's wife syndrome'. Zweig perhaps both left his past behind, but could not stop looking back at it. Zweig certainly looked backed over his shoulder at his own past (rather than anybody else's, and who doesn't?), but he was prepared to leave his own past behind seemingly without regret or longing. For him the art of being everywhere at home consisted precisely in being able to leave behind oneself a good portion of one's past in the place where one had been at home (just as he left behind his home in Salzburg and Benjamin left his past in Berlin behind too). Stone (2011, p. viii) comments that Zweig was 'still at home, but also on the move' and that travel for Zweig was 'a home in itself'. The same could also be said about Benjamin. Being at home everywhere and on the move all of the time is more like the Angels of History and Geography and less like Lot's wife. So much for Zweig suffering from 'Lot's wife syndrome'.

Despite these and some other similarities, Benjamin and Zweig had a profoundly different relationship to theology and to sacred and secular time. Benjamin retained many of the concepts of theology, as Scholem notes and as Benjamin (1999a, p. 471; see also Richter (2016) and Dickinson and Symons, (2016)) himself proclaimed in his famous comment that 'my thinking is related to theology as blotting pad is related to ink. It is saturated with it. Were one to go by the blotter, however; nothing of what is written would remain.'

By contrast, there seems to be little trace of theology in Zweig's writings as he seems to have been a thoroughly secularized Jew and primarily a German-speaking and -writing Austrian, whereas Benjamin was what Brian Britt (2016, p. 11; see also pp. 10, 157, 160, 17) calls 'postsecular' who 'insisted on being fully Jewish and German'. For Scholem (2012, p. 187), Benjamin's 'insights are those of a theologian marooned in the realm of the profane' and who 'translates them into the language of historical materialism'. Yet Benjamin, marooned like the survivor of a shipwreck in the realm of the profane and secular on the desert island of high modernity in the late nineteenth and early twentieth centuries, still found a home in the realm of the sacred (and not just in theology) and in the language of historical materialism, perhaps unlikely bedfellows, but ones which he brought together successfully in a new language of materiality and spirituality.

Scholem (2012, p. 197) goes on later to note the presence in the general of 'the messianic idea' among 'the Jewish categories' that Benjamin introduced into his writings and 'upheld to the last'. Scholem (2012, pp. 101–126) does not specifically mention here (though he does in the following essay in the same collection devoted to 'Walter Benjamin and his Angel') the theological category of the messianic now, or *Jeztzeit.* This is a concept that Benjamin (2003, p. 395) developed coming out of traditional Judeo-Christian theology (with its concept of *Kairos* derived from Greek thought and probably via the Christian apostle and New Testament epistolary writer, St Paul) and combined with Marxist eschatology to indicate a messianic irruption in the present that blasts open the continuum of history (in a word, revolution) (see Eagleton, 1981, pp. 10–11, 73, 146 and 156).

In Benjamin's (1996, p. 483) own terms, in order 'to turn the threatening future into a fulfilled "now", the only desirable telepathic miracle, is a work of bodily presence of mind'. This miracle is performed by the Angel of Corporeality. For Benjamin the body is, or can be, in the fulfilled present in the mind and in the now going into the future. This was certainly the case for Benjamin (1996, p. 483) as he goes on to suggest that 'primitive epochs, when such demeanor was part of man's daily husbandry, provided him, in the naked body, with the most reliable instrument of divination [. . .] What would have become a portent of disaster he binds bodily to the moment, making himself the factotum of his body.' 'Primitive man' does all kind of work *for* his body (dance, ritual, tai chi, yoga, etc.), whereas 'modern man' does all kinds of work *with*, or *on*, his or her body (body-building, diet, exercise, etc.). The traditional human being works his or her whole body and not just parts of it. Rather than the body being a portent of disaster in the future (of disease, decay and death), the body sutures him/her to the present and fulfils the now and the here.

Benjamin later commented on 'modern man' and his relationship to his own body. Writing on the tenth anniversary of Kafka's death, Benjamin (1999b, p. 806) argues that in Kafka's *The Castle*, 'just as K lives in the village on Castle Hill, modern man lives in his body; the body slips away from him, is hostile towards him [. . .] Exile – his exile – has gained control over him[. . .] the most forgotten alien land is one's own body'. Benjamin and Zweig lived this exile from their own bodies, and a double exile from their own homeland. By contrast, for traditional human beings the most fondly remembered homeland is one's own body. Exile is not just a matter of leaving one's home and homeland, but is also experienced in relation to one's own body, something that Benjamin and Zweig wrestled with their whole lives. K is not at home in his own body; the castle is not a home. Being at home includes being at home in one's own body and place. For traditional humans, one's land is one's own body, and vice versa (see Giblett, 2008). They are devotees and recipients of the Angels of Corporeality, Geography and History.

Home is here

An exemplary figure for finding his home here and being at home in his own body and home place, and for writing about both, is Henry David Thoreau. He can perhaps be regarded as a traditional human being with a cyclical sense of time living in the modern age who presaged the postmodern.[1] In his journal entry for 16th July 1851, written four days after his 34th birthday, he recalled 'in youth, before I lost any of my senses, I can remember that I was all alive, and inhabited my own body with inexpressible satisfaction; both its weariness and its refreshment were sweet to me' (Thoreau, 1962, *II*, pp. 306–307). Thoreau looks back with a tinge of longing and regret for the bodily presence of mind he had in his youth. Nearly nine years earlier when he was 24 years old he had already expressed an exilic experience of his own body when he wrote in his journal entry for 21st February 1842 that 'I must confess there is nothing so strange to me as my own body. I love any other piece of nature, almost, better' (Thoreau, 1962, *I*, p. 321).

Thoreau not only looked back to the past to the home of his own body, but also looked in the present to home in places and the future of places. In his youth he was at home in his own body, whereas in his maturity he was not as at home in his body as he was in his youth, but he was more at home in his own country, his homeland. As he grew up and grew older his home changed from his body to his places. Several years later he wrote in his journal entry for 2nd November 1858 how:

> Here I am at home. In the bare and bleached crust of the earth I recognize my friend [. . .] the constant endeavour should be to get nearer and nearer *here* [. . .] A man dwells in his native valley like a corolla in its calyx, like an acorn in its cup. *Here*, of course, is all that you love, all that you expect, all that you are. Here is your bride elect, as close to you as she can be got. Here is all the best and all the worst you can imagine.
>
> (Thoreau, 1962, *XI*, p. 275)

Thoreau's inhabited not only his own body in his youth, but also his own home of his country in his maturity. His habitat was only his own body, but also what is now called his bioregion (see Giblett, 2011, pp. 242–246). Although Benjamin and Zweig, as thoroughly urbanized modern men and wayfaring exiles, did not enjoy this greater ambit of country inhabitation that Thoreau enjoyed living and walking on the urban fringe in the swamps and woods and rowing and sailing on the rivers around Concord in Massachusetts, they did live in various cities as a larger place of home.

Home for Thoreau is variously the body and place in the present. Hope for the future for Thoreau lay in local, uncultivated places, in the country and not in the city. In his most famous essay, 'Walking', he announced that:

> hope and the future for me are not in lawns and cultivated fields, not in towns and cities, but in the impervious and quaking swamps [. . .] I derive more of my subsistence from the swamps which surround my native town than from the cultivated gardens in the village [. . .] When I would recreate myself, I seek the darkest woods the thickest and most interminable and, to the citizen, most dismal, swamp. I enter a swamp as a sacred place, – a *sanctum sanctorum* [a holy of holies]. There is the strength, the marrow of Nature [. . .] A town is saved, not more by the righteous men in it than by the woods and swamps that surround it.
>
> (Thoreau, 1991, pp. 98 and 100)

Thoreau loved being immersed bodily in rivers and swamps with his senses all alive.

He was what could be called 'the patron saint of swamps' and the avatar of the quaking zone (see Giblett, 1996, pp. 229–239; 2009, chapter 1). He is a high priest of earthly sacrality who enters bodily and spiritually into the holy of holies of the temple of the swamp. The swamp is a sacred place set aside in time, space and in the body. Writing in his journal on 30th August 1856 Thoreau maintained that:

> it is in vain to dream of a wildness distant from ourselves. There is none such. It is the bog in our brain and bowels, the primitive vigor of Nature in us, that inspires that dream.
>
> (Thoreau, 1962, *IX*, p. 43)

Bogs and swamps are where the Angels of History, Geography and Corporeality live.

Just as the two million words of Thoreau's *Journal* are the expression of his sense of place of Concord and its surrounds as his home habitat for most of his life, so the thousand pages of Benjamin's *Arcades Project* are the expression of his sense of Paris and its history and culture as his home habitat for part of his life. His earlier writings about Berlin, such as *One-Way Street, Berlin Chronicle* and *A Berlin Childhood Around 1900*, are about the city as habitat and more about his inhabitation of the city than just the various apartments in which he lived and which he visited with their cluttered interiors. Both Benjamin and Thoreau are exemplary expressers of a larger sense of home habitat, which they inhabit bodily and mentally, whether it is in city or country, materially and spiritually. The country and the city, the marsh and the metropolis, are both valuable and have equal claims to attention and care as habitats vital for life on earth. Both Benjamin and Thoreau also had a similar appreciation for swamps, in Benjamin's case for the primal swamps of the pre-historical and pre-patriarchal or matrifocal world and in Thoreau's case for the primal swamps of his geographical world. [2]

The modern man living in the castle of his own body culminates in the Fascist, Benjamin's and Zweig's contemporary, who still lives on. The body is the vehicle and vector by which Fascism sends its message to us in the present and ensures its own survival. The Fascists constituted the body as the communication and transportation technology – the vehicle – by which to communicate their ideology through time and across space (see Giblett, 2008, pp. 90–91). For the counter- or anti-Fascist, such as Benjamin and Zweig, their own writing is the vehicle and vector by which they send their message to us in the present and ensure its survival for the future. In Zweig's case his writing meant his books, whereas for Benjamin his writing was the few books he authored, his scattered essays and reviews, and his precious manuscripts, most of which survived. Also in Zweig's case, art and politics are separate categories (and theology does not enter into it), whereas for Benjamin, as Scholem (2012, p. 185) puts it, 'aesthetic ideas are most intimately bound up with theological categories', which are, in turn as Scholem later says, translated into the political categories of historical materialism. Not so with Zweig.

Dialectical image

One idea in Benjamin's work that runs the full gamut of aesthetics, theology and historical materialism is what he calls 'a dialectical image'. In her relationship to time and eternity, the Angel of History is what Benjamin (1999a, p. 462) called a dialectical image between 'the what-has-been' and 'the fulfilled now'. For him, 'while the relation of the present to the past is a purely temporal, continuous one, the relation to what-has-been to the now is dialectical: is not progression but image, suddenly emergent' (Benjamin, 1999a, p. 462). The 'what-has-been' is not merely (in) the past, but also what has been lost, what is absent or in oblivion, not only past in time, but lost in space, in a place, not only historical, but also geographical, such as wetlands. Benjamin (1999a, p. 464) elaborates further that 'in the dialectical image, what has been within a particular epoch is always, simultaneously, "what has been from time immemorial"'. Wetlands in general have been from time immemorial when the world was wetland. The Angel of Geography is a dialectical image between the what-has-been and the fulfilled here.

'The dialectical image' for Benjamin (1999a, p. 474) is 'the primal phenomenon of history'. It is also the primal phenomenon of geography and corporeality. One such dialectical image of the primal phenomenon of history Benjamin develops is drawn from J. J. Bachofen's *Mother Right*, a theological work in some sense as it is about the pagan divinity of the Great Goddess of the swamps.[3] Benjamin (1996, pp. 426–427; 2002, pp. 11–24) discusses Bachofen directly and extensively in an essay and briefly in a review of an Italian book about Bachofen, without mentioning in either piece Bachofen's discussion of the swamp world. However, in his essay on the tenth anniversary of Kafka's death, Benjamin (1999b, p. 808) reads Kafka's work in Bachofen's terms of the image of 'the swamp world' and translates it into the language of historical materialism. The swamp world is not only a primal phenomenon of history, but also of geography and corporeality, the grotesque lower earthly and bodily strata (see Giblett, 1996, chapter 6; 2008, chapters 4 and 5). Rather than aesthetic sublimation into earthly transcendence (as for Zweig), the swamp world desublimates the earthly and transcendent into slimy and bodily immanence (see Giblett, 1996, especially chapter 2). This is the language of geographical, pre-historical and corporeal materialism.

Benjamin (1999b, p. 809) comments on Bachofen that 'the fact that this stage [of 'the swamp world'] is now forgotten does not mean that it does not extend into the present. On the contrary: it is present by virtue of its very oblivion'. The past is not dead; the past is resurrected in the present, the absent brought back to presence, in and by the dialectical image. The 'what-has-been' of the lost swamp world extends into the present by virtue of its very oblivion.[4] The 'what-has-been' of the lost swamp is not only in the past, in history, forgotten in and by the present, but also in the present, in geography, lost in the city now, present by virtue of its very oblivion.

Sin city

Benjamin looks to the dialectical relationship between past and present in messianic *Kairos*, whereas Zweig looks back at the past, and to the present and future in chronological succession. Given that Lot's wife looked back at the destruction of the two cities of Sodom and Gomorrah, what does Zweig (as suffering from 'Lot's wife syndrome' in Prochnik's view) look at when 'looking back over his shoulder'? More precisely, which cities and their destruction does Zweig, reprising for Prochnik the role of Lot's wife, look back at over his shoulder? Every city, for Harries (2007, p. 78), 'is, potentially, Sodom'. Loader (1990, p. 126) comments that Sodom is 'the most sinful city of all'. Salzburg and Vienna are certainly cities that Zweig looked back to with nostalgia (and Vienna for Zweig had its sinfulness), though he does not look back at their physical destruction as both cities were not physically destroyed by allied bombing after occupation by the Nazis, but both were destroyed politically and culturally by the Nazis, which probably amounted to the same thing as physical destruction for Zweig and to which he 'could not stop looking back [at] over his shoulder'. Yet unlike Sodom and Gomorrah, which were physically destroyed by God for their sins, Salzburg and Vienna were destroyed politically and culturally by the Nazis for what they saw as their sins as cities, in their case of being openly Semitic, culturally free, highly creative and decadently modern.

Other cities, such as Paris, for Zweig were also destroyed culturally and politically (as recounted in *The World of Yesterday*). For Zweig, Paris is 'the city of eternal youth' in a chapter of this title. Ancient and modern Paris is the city of eternal youth and age (though he does not mention its longevity and prehistory), rather than the city of eternal age that ancient Rome was, or aspired to be. Paris for Zweig (2104, p. 149) is tinged in hindsight with longing and regret for 'the wonderfully lively and invigorating Paris of my youth [that] no longer exists [. . .] now that it has felt the iron brand forcibly imprinted on it by the hardest hand on earth'. Zweig was writing when 'German armies and [. . .] tanks were rolling in, like a swarm of grey termites, to destroy' culturally the city of his youth, which lends an ironic and nostalgic tinge to the title of the chapter in which Paris ends up not being quite so eternal and being mortal. Zweig's invocation of a swarm is a trace of theology as the book of *Leviticus* (11: 41) interdicts swarming creatures as an abomination (as we have seen in previous chapters). Zweig follows suit. Zweig certainly looked back over his shoulder in later life at the cultural and political destruction of the Paris of his youth.

Some cities, such as Dresden, Hamburg and Würtzburg, were largely destroyed physically during World War II, the destruction of which neither Benjamin nor Zweig lived long enough to know about. Aerial bombing was seen from its beginnings in the early twentieth century in terms of the biblical story of Sodom and Gomorrah, though not in terms of the story of Lot's wife. If it had, there may have been more reluctance to look (back) at it. Beginning with the first aerial bombing by Italians of Turkish troops in 1911 in North Africa, 'across Europe the popular imagination', according to Lowe (2007, p. 42), was 'fired with images of a man-made Sodom and Gomorrah'. This differed from the divine destruction of the two cities in the biblical book of *Genesis*, but how it differed is not clear except in terms of the agent of destruction. What was certain is that 'man' had become God by taking on a divine function and power of destruction (just as earlier 'he' had taken on the divine function and power of creation in developing cities).

The story of the destruction of Sodom and Gomorrah has for Loader (1990, pp. 38 and 43) 'an anti-urban tendency' and 'bears evidence of an anti-urban attitude' in which 'the sinfulness of life in the city is contrasted with the rustic virtue of hospitality found in nomadic society'. Lot, 'showing as he does respect of the nomadic ideals, is presented as a tragic illustration that the nomad way of life is not tolerated in the city'. Lot's wife, showing as she does disrespect for God's interdiction not to look back at the destruction of the cities, is presented as a tragic illustration that God does not tolerate disobedience. God wins (and Lot and the nomadic way of life win too); Lot's wife loses (and sinful cities lose too). God uses the power of nature, such as seismic activity and chemical reaction, to destroy the cities. The cities were overturned, which for Loader (1990, p. 41) 'points to an earthquake', and 'the sulphurous rain [that descended on them points] to volcanic activity'.

The deployment of what Virilio (2005, p. 15) calls 'our aero-naval modernity' against cities entailed what he also called 'the 'aeropolitics' of a mass extermination of cities' (Virilio, 2007, p. 8). This mass extermination of cities destroyed buildings and their inhabitants, a kind of genocide of cities, a citi-cide, or what Helphand (2006, p. 246) calls 'urbacide', 'the wilful destruction of cities'. This mass extermination of cities occurred at the tail end of World War II. For Virilio (2005, p15), 'after Dresden, and especially after Hiroshima and Nagasaki, this 'aeropolitics' turned into a cosmopolitics of nuclear terror with the *Anti-City* strategy'. The anti-city strategy of World War II was carried out by the fire-storming of strategic bombing.

The images of the divine destruction of Sodom and Gomorrah by 'fire and brimstone' (*Genesis* 19: 24 (King James/Authorized Version)) were pressed into the service of the potential bombing and destruction of British cities in the preamble to World War II. By the 1930s, as Lowe (2007, p. 47) puts it, 'the most enthusiastic champion of bombing' in England was, hardly surprisingly, 'forced to admit to a widespread pessimism among his contemporaries, who foresaw "a fate comparable to that of Sodom and Gomorrah" for British cities'. The fate of Sodom and Gomorrah was mainly to befall German cities as pre-emptive strikes designed to avoid the repetition of the 'man-made' destruction of Sodom and Gomorrah with British cities. Yet the results for Lowe (2007, p. 48) were 'no longer a vision of Sodom and Gomorrah', but 'a vision of Armageddon', the battle culminating in and at the end of the world. Similar sentiments were voiced in Nossack's (2004) witness-survivor account of the bombing and destruction of Hamburg simply entitled *The End: Hamburg 1943*.

Rather than Armageddon, the bombing of Hamburg in 1943 was called 'Operation Gomorrah', as noted by Harries (2007, p. 84) and Sebald (2003, p. 26). Sebald (2003, p. 57) later uses 'Operation Gomorrah' as a shorthand, non-euphemistic term for the bombing of Hamburg with all its attendant horrors and terrors. He also later invokes Benjamin's (2003, p. 392) 'Angel of History' looking at piles of wreckage and debris not unlike that produced by 'Operation Gomorrah' (Sebald, 2003, pp. 67–68). In *Terror from the Air* Sloterdijk (2009, p. 103) relates (and cites) how pre-war anti-establishment Viennese journalist Karl Kraus in 1936 in the last issue of *Die Fackel* published four months before his death gave up his fight against what he called 'the air of Sodom'. Kraus sensed that the oxygen had been sucked figuratively, culturally and politically out of the life-giving air of pre-war Europe into a fiery conflagration of destruction nearly a decade before this occurred physically in fact. Perhaps he also prophetically or presciently foresaw aerial firebombing in World War II that sucked oxygen out of the air literally. On a post-war tour of Europe, Stephen Spender (cited by Harries (2007, p. 132, n.24)) commented that 'we can regard Germany as Sodom'. For Kraus it was already was Sodom before the war, and not just because it was sinful. The pre-war atmosphere of Europe was already toxic; bombing only enacted it and brought it to fruition. All these writers who invoke Sodom or Gomorrah do not mention Lot's wife, as if they were afraid of being turned into pillars of salt themselves for looking back at the destruction of German cities or European culture.

Some historians of the past seem to be unable to take their own advice, or are at least unable to be reflexive about their own position and role looking back into the past. As the epigraph to the chapter on survival in his book about the bombing of Hamburg, Lowe (2007, p. 277) cites the warning of *Genesis* 19:17: 'Look not behind thee [. . .] lest thou be consumed'. He does not follow this advice, nor heed this warning himself. Perhaps he does not think it applies to him but only to the survivors of the bombing of Hamburg. He explains in an endnote at this point that these are 'the words of the angels to Lot, as they saved him from the destruction of Sodom and Gomorrah. Lot's wife did look back and, as a consequence, was turned into a pillar of salt', as *Genesis* 19: 26 goes on to relate.[5] Looking back has both a spatial and temporal dimension and inflection in retrospection. Just as prospect has both a spatial and temporal dimension of looking over a place or forward into time, so is retrospect as it involves looking back into the past and at a lost place; just as prospects can be either pleasing or displeasing spatially and temporally, so can retrospects. Lot's wife not only looks back at the place that has been destroyed, but also tries to look back into the past that has also been destroyed. For this double sin she was turned into a pillar of salt.

By contrast, for a witness-survivor of the destruction of Hamburg such as Erich Nossack (2004, p. 23) 'we no longer have a past', whereas the historians of destruction still have a past, or the past. The historian looks back into the past of other people, and tells others about it. The survivor who no longer has a past is compelled to 'bear witness' as Nossack (2004, p. 2) puts it of the loss of the past and 'to render an account' otherwise it will 'gradually fade like an evil dream'. The witness gives an account in the present of the past that is lost and does so for the future. The witness is writing for the future whereas the historian writes for the present. In this respect, the witness is more like the Angel of History than the historian who has a past, albeit of other people.

Hamburg was 'the first big city to be annihilated' according to Nossack (2004, p. 32), who witnessed its destruction as a spectator from outside the city (pp. 1–2) and bore witness to it in *The End*. The title refers not only to the end of the city as he knew it, but also the end of the world in apocalyptic tones. What he calls 'the ruin of Hamburg' (p. 6) is also variously described in apocalyptic terms as 'the abyss' (p. 6), 'the end' (p. 8) and 'the netherworld' (p. 11). Not merely 'anti-city', the destruction of Hamburg also pitted nature against nature for 'even nature had risen up in hatred against herself' (p. 11). Aero-naval modernity is not only anti-city but also anti-nature in which two forces of nature (fire, storm) are both harnessed by the militaristic men and unleashed against nature. The result for Nossack is 'the raging of the world against itself' and 'the earth writhed in agony' (p. 15). These are the death throes of 'raging destruction' (p. 13) inflicted by the world and the earth upon itself by a world- and earth-hating military whose end (in two senses) is 'the ruined world' (p. 13). 'The horror' of 'the disaster' (p. 16) was that it was both natural and cultural, the forces of nature directed against both nature and culture by natural and cultural human beings against other natural and cultural human beings.

Part of the horror for Nossack (2004, p. 17) was the fact that the refugees fleeing from the city 'brought with them an uncanny silence. No one dared question these mute figures.' They had ceased to be human, to be natural and cultural beings, because they were deprived of speech and had become speechless automatons who mimicked and mocked human behaviour in uncanny fashion. They were exiled many times over: from their home place, their own history and their own bodies, and the Angels of all three. The fact that this disastrous destruction of Hamburg was 'the end' for its residents of 1943 meant that not only 'we no longer have a past' (p. 23), but also that they only had a present and future rent from any connection with the past. Perhaps they did not even have a present and future as 'we no longer had any time at all, we were outside of time' (p. 29). In other words, everything was 'denuded of time' and 'had become eternal' (p. 40). Space and time had ceased to exist as they had been known.

Not only was the spatial landscape transformed, but also the temporal one as well: 'the infinite behind man wafted unhindered in the endlessness before him and hallowed his countenance for the passage of what is beyond time' (p. 29). Space and time are collapsed together into the landscape of eternity and infinity. The less, or more, than human figures, who people this landscape are not only mute automatons but also icons with eyes 'grown larger and transparent' (p. 29) through witnessing the horrors of fiery destruction of their city and the (its/their) past. They had become lifeless images of the Angel of Corporeality.

The Angel of Geography looks over the destruction of places in the past, present and future and bears witness to it in a dialectical image between the what-has-been and the fulfilled here. The Angel of History looks back at the acts and events of destruction in the past, but moves forward into the future, as does the historian of the present. How the Angel of History, the historian of the past and the historian of the present relate to the past, present and future are different. For Herman Knell (2003, p. 2) who witnessed and survived the destruction of Würtzburg and wrote about it as an historian of the present, 'there is a psychological need to forget and a moral obligation to remember. There is the human desire to forgive and the ethical necessity to warn of a possible repeat of the disaster.' The survivor-witness is suspended between a past that they may wish to forget, but must remember in the present and for the future in order to try to avoid the possible repetition of the disasters of the past in the present and future in every place.

Notes

1 For Thoreau's cyclical sense of time, see 'The Seasons' in Giblett (2013, pp. 149–163).
2 For Thoreau's views of the city and the swamp (which were not advanced merely *for* the swamp and *against* the city) see Giblett (2016, pp. 171–181).
3 Recent discussions of Benjamin and theology, such as Britt (2016), and Dickinson and Symons, eds (2016), do not consider Bachofen at all, perhaps because it is neither Christian, nor Jewish, and/or perhaps because theology for them is about God and not about the Great Goddess. Perhaps she should be considered in a discussion of Benjamin and 'ge-ology'.
4 Elsewhere I have employed Benjamin's pronouncement as a motto for the relationship between many cities to their geographically lost and historically forgotten wetlands; see Giblett (2016, especially pp. 50–51 in relation to Paris, or *Lutetia*, 'the filthy marsh').
5 It is noteworthy that Lowe (2007, p. 403, n.1) relegates Sodom and Lot's wife to a note here and that Harries (2007, p. 132, n.24) also relegates 'Germany as Sodom' to a note.

References

Benjamin, W. 1996. *Selected Writings, Volume 1: 1913–1926*. M. Bullock and M. Jennings, eds. Cambridge, MA: Belknap Press of Harvard University Press.
Benjamin, W. 1999a. *The Arcades Project*. H. Eiland, trans. Cambridge, MA: Belknap Press of Harvard University Press.
Benjamin, W. 1999b. *Selected Writings, Volume 2: 1927–1934*. M. Jennings, H. Eiland and G. Smiths, eds. Cambridge, MA: Belknap Press of Harvard University Press.
Benjamin, W. 2002. *Selected Writings, Volume 3: 1935–1938*. H. Eiland and M. Jennings, eds. Cambridge, MA: The Belknap Press of Harvard University Press.
Benjamin, W. 2003. *Selected Writings, Volume 4: 1938–1940*. H. Eiland and M. Jennings, eds. Cambridge, MA: The Belknap Press of Harvard University Press.
Britt, B. 2016. *Postsecular Benjamin: Agency and Tradition*. Evanston, ILL: Northwestern University Press.
Dickinson, C. and Symons, S. eds 2016. *Walter Benjamin and Theology*. New York: Fordham University Press.
Eagleton, T. 1981. *Walter Benjamin, or Towards a Revolutionary Criticism*. London: Verso.
Foucault, M. 1977. *Discipline and Punish: The Birth of the Prison*. A. Sheridan, trans. Harmondsworth: Penguin.
Freud, S. 2003. *The Uncanny*. D. Mclintock, trans. London: Penguin. First published 1919.
Giblett, R. 1996. *Postmodern Wetlands: Culture, History, Ecology*. Edinburgh: Edinburgh University Press.
Giblett, R. 2008. *The Body of Nature and Culture*. Basingstoke: Palgrave Macmillan.
Giblett, R. 2009. *Landscapes of Culture and Nature*. Basingstoke: Palgrave Macmillan.
Giblett, R. 2011. *People and Places of Nature and Culture*. Bristol: Intellect Books.
Giblett, R. 2013. *Black Swan Lake: Life of a Wetland*. Bristol: Intellect Books.

Giblett, R. 2016. *Cities and Wetlands: The Return of the Repressed in Nature and Culture*. London: Bloomsbury.

Harries, M. 2007. *Forgetting Lot's Wife: On Destructive Spectatorship*. New York: Fordham University Press.

Helphand, K. 2006. *Defiant Gardens: Making Gardens in Wartime*. San Antonio: Trinity University Press.

Knell, H. 2003. *To Destroy a City: Strategic Bombing and its Human Consequences in World War II*. Cambridge, MA: Da Capo.

Loader, J. A. 1990. *A Tale of Two Cities: Sodom and Gomorrah in the Old Testament, Early Jewish and Early Christian Traditions*. Kampen: J. H. Kok.

Lowe, K. 2007. *Inferno: The Fiery Destruction of Hamburg, 1943*. New York: Scribner.

Nossack, H. 2004. *The End: Hamburg 1943*. Joel Agee, trans. Chicago: University of Chicago Press.

Prochnik, G. 2014. *The Impossible Exile: Stefan Zweig at the End of the World*. London: Granta.

Richter, G. 2016. Benjamin's Blotting Paper: Writing and Erasing a Theological Figure of Thought. *In: Inheriting Walter Benjamin*, London: Bloomsbury, pp. 35–57.

Scholem, G. 2012. *On Jews and Judaism in Crisis: Selected Essays*. W. Dannhauser, trans. Philadelphia: Paul Dry.

Sebald, W. 2003. Air War and Literature. *In: On the Natural History of Destruction*, A. Bell, trans. New York: Random House, pp. 1–104.

Sloterdijk, P. 2009. *Terror from the Air*. A. Patton and S. Corcoran, trans. Los Angeles: Semiotext(e).

Stone, W. 2011. Introduction. *In*: S. Zweig, *Journeys*, London: Hesperus, pp. vii–xv.

Thoreau, H. 1962. *The Journal 1837–1861, Volumes I–XIV*. B. Torrey and F. Allen, eds. New York: Dover.

Thoreau, H. 1991. *Walking*. Boston: Beacon. First published 1862.

Virilio, P. 2005. *City of Panic*. J. Rose, trans. Oxford: Berg.

Virilio, P. 2007. *Art as Far as the Eye Can See*. J. Rose, trans. Oxford: Berg.

Weidermann, V. 2016. *Summer before the Dark: Stefan Zweig and Joseph Roth, Ostend 193*. C. Janeway, trans. London: Pushkin Press.

Zweig, S. 2014. *The World of Yesterday: Memoirs of a European*. A. Bell, trans. London: Pushkin Press.

Zweig, S. 2016. *Messages from a Lost World: Europe on the Brink*, translated by Will Stone, London: Pushkin.

9 Rainbow Serpent anthropology, or Rainbow Spirit theology, or swamp serpent sacrality and marsh monster maternity?

The Rainbow Serpent is a well-recognized figure in Australian Aboriginal culture, or at least in white fellas' study of it, though there is no unitary or homogeneous Australian Aboriginal culture, but diverse and heterogeneous Australian Aboriginal culture*s*. This is certainly the case with the Rainbow Serpent. While many Australian Aboriginal cultures have a water being in their ongoing creation or dreaming stories, this being is not necessarily rainbow-coloured. In addition, the idea of 'the Rainbow Serpent' supposedly common to many Australian Aboriginal cultures is an invention or generalization by anthropologists from stories told and beliefs held in 'widely separated parts of Australia', as one anthropologist put it.

Similarly 'the Rainbow Spirit' is an invention or creation by Australian theologians (Aboriginal and non-Aboriginal) who seem to want to deflect the negative biblical connotations of Satan the serpent in the book of *Genesis* as an evil figure, and of water beings in the Bible as evil figures, by sublimating or transubstantiating the Rainbow Serpent into the Rainbow Spirit. Despite its problematic nature, 'the Rainbow Spirit' and its theology comes out of a worthwhile desire to be and act as a creative figure engaged in cultural dialogue between black and white peoples, Aboriginal and non-Aboriginal Australians, about caring for country or the land from within a cross-cultural ecotheology. I am not an Aboriginal person so I do not speak, or claim to speak, with any cultural authority about Aboriginal cultures. I rely on published sources. I am just a white-fella trying to learn from and, engage in dialogue with, Aboriginal peoples via the written word and to share that learning with non-Aboriginal peoples.

Jasmine Corowa's colourful and exquisite illustrations of the Rainbow Spirit in the written texts devoted to Rainbow Spirit theology depict it as a snake or serpent (The Rainbow Spirit Elders, 2000; 2007). In the Rainbow Spirit literature the association of the rainbow spirit/snake/serpent with the evil serpent or snake of the biblical story of the Fall is acknowledged briefly and the relationship with the Rainbow Serpent is also discussed briefly. In the present chapter I do both more extensively by introducing the Rainbow Serpent and the anthropological literature around it, then moving on to discuss the Rainbow Spirit and the theological literature around it. This chapter proposes a rapprochement between the two figures – serpent and spirit – and between the three -ologies in order to produce a dialogic anthropo/eco/theo/logy, or even better sacrality, between spirituality and materiality that cares for earth.

Rather than these -ologies that privilege the logos or word and the -isms in general that exploit and destroy earth (industrialism, capitalism, neo-liberalism, etc.), and sanctuarism in particular that only cares for special places while the rest of earth goes to hell in a hand-basket, the -ities in general and sacrality in particular care for earth and all earthly places (for the '-isms' and '-ities', see Giblett (2011, especially Figure 2, pp. 32–34)). This chapter (and the book) concludes by arguing for marsh monsters and swamp serpents as figures that embody and express this earthly sacrality in the Symbiocence. They, and earth, should be treated with respect and reverence by living bio- and psycho-symbiotic livelihoods in bioregional home habitats of the living earth where they and all living creatures live.

The politics and practices of sacrality and symbiosis are exemplified today in Australian Aboriginal Country. Here there is no landscape to be culturally constructed or greedily consumed, no discourse to produce an object and a subject, but a loving union of people and place, culture and country. As Hugh Webb eloquently and powerfully puts it:

> there is no emotional or intellectual need to stand *outside of* the land (wet or dry) to see it as something to be regarded aesthetically, something to be developed, something to be tamed. For lands are the sites – in some senses, the 'meanings' – of very important cultural stories.
>
> (Webb, 1996, p. 5; his emphasis)

Aboriginal Country is not a construction, so it cannot be deconstructed. It is not a discourse, an institutionalized way of seeing, saying and doing, but a symbiosis, an intimate and intuitive way of being. Aboriginal Country is outside history (white chronological history); it is (in) prehistory, it is (in) the Dreaming; it is not in the past, but in the past, present and future, the now, the nowtime. It precedes and refuses the colonization of time. It is what could be called a 'natural cultural landscape' to oppose the 'cultured natural landscapes' of the gentleman's park and enclave estate, national park, wilderness, mining, pastoralism, the bush, and the city (see Giblett, 2011).

Rainbow Serpent

Josephine Flood (1995, p. 171) in her classic book, *The Archaeology of the Dreamtime: The Story of Prehistoric Australia and its People*, notes that 'the appearance of the Rainbow Serpent belongs to the years between 9,000 and 7,000 years ago which would make the Rainbow Serpent myth the longest continuing religious belief documented in the world'. It comes out of the longest continuing cultures in the world dating from at least 80,000 years ago.

As an aside, the order of priority in the sub-title of Flood's book renders indigenous people a subset of, object of ownership by and appendage to the land or country or nation and writes 'native title' out of history. Why not *The Story of Prehistoric Australian People (and Their Lands)*? The land is the people's, and the people are the land's. The concept of 'the Dreamtime' is also problematic because, as the Rainbow Spirit Elders (2007, pp. xi and 38) put, 'it suggests that [a spiritual dimension of] reality only existed at the time of creation; we Aboriginal people know that this reality continues to exist in the present'. The term 'Dreaming' is preferred, referring to 'Dreaming sites, stories and totems'.

This 'myth' or 'religious belief' was first noted formally by an anthropologist in 1926 when A. R. Radcliffe-Brown (1926, p. 19) announced to his fellow anthropologists that 'there is found in widely separated parts of Australia a belief in a huge serpent which lives in certain pools or water-holes. This serpent is associated, and sometimes identified, with the rainbow.' This serpent could also be regarded as a 'dreaded monster' who was 'at times tricky and malignant. But he could do a good turn to men already possessed of some magical quality' (Radcliffe-Brown, 1926, p. 20). In other parts of Australia, Radcliffe-Brown (1926, p. 21) noted that the serpent 'swallows human beings whole' and is even in one place 'endowed not with huge teeth only, but also with a special craving for the black-fellow'.

The Serpent could be regarded as an orally sadistic monster and so a bringer of death. But the Serpent is the protector of water-holes and so of water with its life-giving properties. Indeed, the 'Rainbow Serpent', Charles Mountford (1978, p. 23) argued, 'is essentially the element of water'. Or, as Radcliffe-Brown (1930, p. 342) argued in a follow-up article on the subject, the Serpent even 'represents the element of water which is of such vital importance to man in all parts of Australia'. Water is both life-giving and death-dealing in Australia as the land 'of droughts and flooding rains', to cite a famous line of Dorothea MacKellar's jingoistic poem, 'I love a sunburnt country'. The Serpent seems to represent simultaneously the powers of life and death, the former of which is so vulnerable and the latter of which is such a constant threat in most of Australia.

As the element of water the 'Rainbow Serpent' cannot be identified with any particular water-hole, but only as inhabiting the water-holes of a region. Such is the case with the Waugal, the 'Rainbow Serpent' in, or of, the wetlands and rivers of south-western Australia where:

> Waugal beliefs are widespread [. . .] and refer to a water-creative force with a serpentine physical manifestation [. . .] The Waugal is not *just* a mythic serpent, an Australian version of the Loch Ness Monster. The Waugal is not *just* a totemic ancestor. The Waugal is not *just* a spiritual being, a semi-deity. The Waugal is indeed all of these but is, more fundamentally, a personification, or perhaps more correctly animalisation, of the vital force of running water. As such, the question 'does this permanent river (or creek, or spring, or other water source) have (or belong to, or be associated with) a Waugal (or the Waugal) becomes, from an Aboriginal view point, meaningless and condescending. The presence of 'living water' bespeaks Waugal immanence.
>
> (O'Connor, Quartermaine and Bodney, 1989, p. 47)

Mircea Eliade (1958, p. 193) has traced how 'living water, the fountains of youth, the water of life [. . .] is guarded by monsters'. And the Waugal is no exception. In the case of Forrestdale Lake, one of the internationally important wetlands in the Perth metropolitan area in Western Australia, O'Connor, Quartermaine and Bodney point out that disturbance of the native reeds (*Baumea articulata*) around the lake's edge is forbidden as this could unleash the Waugal's destructive power. Yet Waugal beliefs are not merely of mythological and anthropological interest as an exotic curio or quaint kind of kid's story. They are indicative of a fundamental cultural difference in the perception and positioning of Western Australian wetlands (see Giblett and Webb, 1996).

One of the iconic tourist lakes around Perth is called Loch Ness, so the association that O'Connor, Quartermaine and Bodney make with that monster is not arbitrary or tenuous. It is also one that Shane Pickett seems to deliberately evoke in one of his paintings of the Waagle, now held by the City of Fremantle in Western Australia (and reproduced as the frontispiece of the present book; see also Pickett, 2004).[1] In this painting, the more than Loch Ness monster of the Waagle rises out of the water against a backdrop of a stratified landscape which could also be the stratigraphic layers of the land so this water could equally be an aquifer, or an underground river, or an overground river, or a rain cloud, thus indicating that for Aboriginal people there is no distinction between waters under, on, or above the earth in the air, in the sky. They are all waters. There is no grotesque lower earthly (and bodily) strata, no netherlands of the slimy swamps and marshes, no monumental upper earthly (and bodily) strata, no upper heights of the surmountainous sublime. There is one living body in which one strata is not privileged over any another, but all strata go to make up, and function as, a living whole.

This painting portrays (it is not a landscape) the earthly sacrality of marsh monsters and swamp serpents as a belief in, and as a way of living, bio- and psycho-symbiotic livelihoods in bioregions of the living earth in the Symbiocene. This painting portrays the Symbiocene as the geological age of the laying and layering up and down of bodies of land, waters, air and living beings in biogenic and non-biogenic strata (discussed in greater detail and differentiated from the Anthropocene below).

In a brilliant discussion of Western Australian Aboriginal attitudes to their rivers and wetlands, Hugh Webb (1996, pp. 67, 70) cites the Seaman inquiry into Aboriginal lands which refers to 'people who have country in the sea'. This led to the assertion by a number of groups that Aboriginal people should be able to own the beds of rivers if they run across or through the surface of their land. Presumably this would also extend to aquifers, or underground rivers, that run *through* (rather than merely under) the depths of their land too. Pickett's painting does not draw a distinction between land and water under ground and that above ground. Land and water are one – contiguous on the surface and in the depths. Janice Lyndon brilliantly illustrates this contiguity and continuity in a recent kids' book about the Wagarl (Little, 2004, p. 6).

By attempting to exclude water-rights from Native Title, former Prime Minister John Howard was not only dealing a cruel and savage blow to reconciliation.

He was also demonstrating his ignorance that water and land cannot really be separated out in this way for both Anglo and indigenous Australian cultures and for the Australian continent (and its shelf). Water is the life-blood of land, and land is people – both Anglo and indigenous. Pickett's painting of the Waagle painted twenty-five years ago and now housed in the Art Gallery of Western Australia illustrates this point by showing the Waagle rising out of a river of blood bearing people on its (or his or her) back (see Pickett, 2003). This painting portrays the Symbiocene as the inter-corporeal relationship between the layers of land, water, air and living beings in which healthy bodies, minds, lands and waters can and do flourish.

A river of blood in European culture and for their settler diasporas, such Anglo-Australians, is typically horrific. It has been associated with menstrual flow, massacre, war, sacrifice, abattoirs, etc. So horrific is this association that menstrual flow was censored from my first published discussion of this topic by the prestigious, yet squeamish, Melbourne University Press (see Giblett, 2007, p. 37), and reinstated in a subsequent publication by the equally prestigious, but more robust Palgrave Macmillan (see Giblett, 2009, p. 185). In pondering the Judeo-Christian significance of blood I referred to my thirty-five-year-old Bible dictionary (as one does) and found that the first comment under the heading of 'blood' is that 'the point chiefly to be determined is whether 'blood' in biblical usage points basically to life or to death'. Like wetlands, blood points to *both* life *and* death. By contrast, biblical binarism and moralism separates life and death, and so constructs blood as associated with *either* life *or* death. Regarding water as both life-giving and death-dealing means that the Waagle is not a monstrous swamp serpent who only kills and consumes, but is both the bringer of death/taker of life, *and* the giver of new life as well.

Rather than rainbow-coloured, Mudrooroo[2] (1996, pp. 33 and 36), maintains that the Waugal of south-western Australia is black and furthermore suggests that the idea of a 'Rainbow Serpent' common to all Australian indigenous cultures is an anthropologist's invention:

> Watjelas [white fellas] have studied us and have found that Aborigines all over Australia respect snakes, and they have joined up all these stories about snakes and made something called a Rainbow Serpent. They say and even tell us that Waugal is a rainbow serpent, whatever that is. But he isn't. He is a big, hairy snake that made the rivers and hills and valleys and then, after he had done this, went to sleep in the deep part of the river. If he is any colour, he is black, but when we tell them this, they say he is a Rainbow Serpent and refuse to listen.

The myth of the 'Rainbow Serpent' could thus be seen as having developed out of a drive to unify and homogenise the heterogeneous, even to imply that there must be some sort of originary and unitary ur-myth (*a la* Casaubon's 'the key to all mythology' in George Eliot's *Middlemarch*) from which contemporary stories are derived, and of which they would be mere vestiges. Such a drive seems to be the product of a quest for origins, rather than the result of a desire to understand the geographical richness and contemporary relevance of Australian Aboriginal cultures.

The living water of the wetland in Aboriginal cultures is worlds away and cultures apart from the dead water of the marsh, swamp, slough, etc., in patriarchal culture. But out of dead water, living water comes and without the latter the former would not be possible. Out of old life, the death and decay of the swamp, new life springs, the blossoms, fruits and foliage of the swamp, and the animals that feed on them. The 'Rainbow Serpent', Radcliffe-Brown (1930, p. 347) concludes, 'may be said to be the most important representation of the creative and destructive power of nature, principally in connection with rain and water'. The 'Rainbow Serpent' is not a monstrous swamp serpent which only kills and consumes, but the Great Mother/Goddess of new life as well. The 'Rainbow Serpent' is, in the words of the Noonuccals (1988, unp.), 'the giver and taker of life'. Swamp waters are both life-giving and death-dealing – living black waters.

The combination of the giving and taking of life represents the mixing of birth and death in the swamp's ecology. Radcliffe-Brown (1926, pp. 23–24) argued earlier that the function of the 'Rainbow Serpent' is 'to maintain at their full power the forces of nature':

> the processes of nature, the changes of the seasons, the growth, flowering, and fruiting of plants, the multiplication of animal species and of the human species itself, are not considered as just happening, but must be produced or provided by the society itself by means of the co-operative efforts of the various totemic groups amongst whom the various realms of nature are distributed. Now the 'Rainbow Serpent', whether regarded as a rainbow or a serpent, or species of serpent that lives in water-holes, is just as much a part of nature as the kangaroo or the sun or cold weather, and is therefore just as necessarily an object of ceremonies as these in any well-organised system.

The 'Rainbow Serpent', as Kenneth Maddock (1978, pp. 2 and 8) put it, 'in the water, but also in the sky' (and in the earth, I would add, hence all the strata in Pickett's painting) is not only 'responsible for female fecundity, but [also] responsible for destructive forces'. Indeed, the 'Rainbow Serpent' combines productive and destructive forces into one force, 'unites opposites in a totality', just as in the swamp's ecology. The 'Rainbow Serpent' is 'both male and female' and 'both benignant and malicious', even malignant. But not exclusively the malignant marsh monsters and swamp serpents of patriarchal and filiarchal Western culture.

Rainbow Spirit theology

In the mid-1990s a group of Australian Aboriginal Christians calling themselves 'the Rainbow Spirit Elders', together with some non-Aboriginal theologians, set out to develop 'an indigenous Aboriginal theology' from the 'starting point [of] the land as a central spiritual reality' (Rainbow Spirit Elders, 2007, pp. vii, viii).

In order to do so and to flesh out this theology, they identified the 'Rainbow Spirit' as a profound and 'universal symbol for an indigenous Aboriginal theology' (p. ix). In particular, as 'the Creator, the Rainbow Spirit gave life to all our ancestors and all the creatures – the trees, plants, animals and birds – and to the landscape itself' (p. 14). In addition, 'our people have been entrusted by the Creator Spirit with the care of the land' as co-creators, caretakers and trustees (p. 5; see also pp. 3 and 35).

The Rainbow Spirit Elders acknowledged right up front the tradition of the Rainbow Serpent within which they were working (without mentioning it per se, or citing any of the anthropological sources discussed above). They also addressed some of the problems the Rainbow Spirit/Serpent might present for traditional Christians steeped in biblical theology. They relate how the Rainbow Spirit as the Creator is:

> Often portrayed by our Aboriginal artists as a powerful snake who emerged from the land, travelled the landscape leaving trails of life, and then returned to the land through caves, waterholes and other sacred sites. Early Christian missionaries associated this snake with Satan and the story of the Fall in *Genesis* 3. They saw the snake as a pagan symbol of fertility religion and condemned it as evil [. . .] For a long time, Christian Aboriginal people have accepted the teaching of the missionaries that the Rainbow Spirit is a symbol of evil. We believe it is time to return to this symbol in our culture to help us rediscover our spiritual identity as Christian Aboriginal people. We believe that the Rainbow Spirit is not the source of evil but of life.
>
> (Rainbow Spirit Elders, 2007, pp. 13–14)

This list of beliefs refuses point-blank and head on the pejorative association between the Rainbow Serpent of creation and the satanic serpent of the Fall may have for Christian Aboriginal people. Yet these beliefs raise a number of questions and problems that simply cannot be ignored or passed over lightly without comment.

For instance, where do these beliefs leave non-Christian Aboriginal people? Could not the Rainbow Spirit/Serpent also help them to rediscover their spiritual identity as Aboriginal people? Why would they need to take on the cultural baggage of Christianity as well (especially given its problematic record in Australia as the elders state)? What about non-Christian non-Aboriginal people too? Could not the Rainbow Spirit/Serpent also help non-Christian, non-Aboriginal people to (re)discover their/our spiritual identity as creatures and carers of the earth? Are not all people entrusted with the care of the land? What is wrong with the snake as 'a pagan symbol of fertility religion'? Can it not be praised and celebrated as good?

The Rainbow Serpent is a figure of creation of life and good, and clearly not the satanic serpent of the Fall who introduced evil and death into the Paradise of the Garden of Eden. Indeed, the elders later point out that 'in many of our traditions the Rainbow Spirit is depicted as a massive snake whose creative powers transform the earth [. . .] The Rainbow Spirit is the life-giving power of the Creator Spirit active in the world' (Rainbow Spirit Elders, 2007, p. 31; see also p. 56). And not the exclusively death-dealing evil serpent of the Fall. Earlier the Rainbow Spirit theologians in their reading of *Genesis* 1 slip from 'the Rainbow Spirit' to 'the Rainbow Snake' (Rainbow Spirit Elders, 2000, commentary on 'Painting One', unp.). The back cover blurb of *The Rainbow Spirit in Creation* equates them. The Rainbow Spirit Elders do not tackle and discuss the figure of the serpent in the story of the Fall in *Genesis* 3 and its pejorative associations. Who or what could embody evil and be the tempter of Eve in the Garden of Eden to eat of the tree of the knowledge of good and evil in disobedience of God's commandment not to do so? If the Rainbow Spirit/Snake is the creator, who is the destroyer? If the black snake is good, who is evil? Perhaps it is white robber barons – industrial (factory owners), pastoral (squatters), mineralogical and metallurgical (mining magnates).

The Rainbow Spirit theologians seem to be perpetuating a number of anthropological and theological 'myths' (unexamined assumptions, accepted truths, everyday verities lived by): first, the myth of the universality of the Rainbow Spirit/Serpent among Aboriginal cultures throughout Australia (when we have seen that the Waugal/Waagle for some is a black serpent and Corowa depicts the Rainbow Spirit/Snake as black in her first three paintings of the *Genesis* creation story (Rainbow Spirit Elders, 2000, unp.)); secondly, the myth of the exceptionalism of a group of God's chosen people with a manifest God-given destiny to save themselves (variously Jews, Christians, Americans, commodity consumers, even the human species[3] and, in this case, Christian Aboriginal people), or the earth, or both; thirdly, the myth of a hard and fast distinction between good and evil, and between the spirit world and the material world with beings in the latter used as moralized symbols of qualities in the former (when for Aboriginal cultures, as one of the Rainbow Spirit Elders points out elsewhere (as discussed below), the two worlds are intertwined, not moralized and the relationship between them not merely symbolical); and fourthly, the myth of a hard and fast distinction between Judeo-Christianity and paganism which many Christian missionaries certainly perpetuated, but which neither the Bible, nor the Christian cultural tradition supports (think of the similarities between Judeo-Christian and non-Christian creation stories, and of Christmas and Easter with their 'pagan' symbols of fertility, such as eggs, trees, rabbits etc.).

The Rainbow Spirit/Serpent is a being that can be used to deconstruct and decolonize the pagan/Christianity divide and the colonization of the former by the latter. Like snakes in general, the Rainbow Spirit/Serpent is not necessarily benign (as we have already seen). The Rainbow Spirit Elders go on to point out that 'the Rainbow Spirit is an important being in our ancient stories of the beginning. In these stories, the Rainbow Spirit swallowed young people and regurgitated them as young adults' (Rainbow Spirit Elders, 2007, p. 14). In other, traditional anthropological words, the Rainbow Spirit/Serpent is an important being involved in initiation rituals and rites of passages from childhood to adulthood. This can be a painful, but necessary process. It is not necessarily good or bad, it is neither good nor bad, but a part of life, of growing up and of becoming a full adult member of a culture with all the pleasures, privileges and responsibilities that that involves.

The material and spiritual worlds are intertwined for one of the Rainbow Spirit Elders who said 'Aboriginal culture is spiritual. I am spiritual. Inside of me is spirit and land, both given to me by the Creator Spirit [. . .] the land, too, is spiritual [. . .] The land owns me' (Rainbow Spirit Elders, 2007, p. 12). Land is not something outside this elder, but inside him or her. He or she is spiritual, like the land. The land is thus not something that can be owned, but the land is the owner of the person. Nor is the Rainbow Spirit a symbol of something spiritual, such as of a spiritual or moral quality, like good (or evil). The Rainbow Spirit is spiritual (as the name implies). The material rainbow (and the serpent for that matter) for the Christian Rainbow Spirit Elders is 'a symbol of our spiritual unity in Christ' (Rainbow Spirit Elders, 2007, p. 7). It can also be a spiritually unifying force in the land for all people (without being a symbol (of anything)). One of the elders claimed in the second edition that *Rainbow Spirit Theology: Towards an Australian Aboriginal Theology* is 'the foundation text for the vision of an Australian spirituality' (Rainbow Spirit Elders, 2007, p. xix). Indeed, it might be the vision of an Australian spirituality for all Aboriginal and non-Aboriginal, Christian and non-Christian Australians alike and be pertinent in other countries for other people too. After all, it is the oldest documented spiritual belief in the world. And many countries and cultures have stories of marsh monsters and swamp serpents (as we have seen in previous chapters). They could also be regarded as figures of reverence and worship.

One aspect of this vision of an Australian spirituality for all people is that the land for Aboriginal people, including the Rainbow Spirit Elders, is not a passive object, but an active agent; not a thing, but a being; not dead, but alive. The land for the Rainbow Spirit Elders is 'alive within [. . .] The land is itself alive, dynamic and creative' (Rainbow Spirit Elders, 2007, pp. 31–32). Patrick Dodson, a noted Aboriginal leader and now a Labour Senator, is quoted as saying that the land 'is not a thing – it is a living entity' (cited p. 32).

The land as living entity can be summed up in one word – mother. The land is mother who gives life out of her womb and who receives death into her womb, out of which new life is created. One of the Rainbow Spirit Elders stated that 'when a person dies, the land, as mother, opens her womb and takes the body. But the spirit is not in the land. Only the shell, the body, stays in the land; the spirit goes to [. . .] the Rainbow Spirit' (Rainbow Spirit Elders, 2007, p. 34). The land is not a tomb, but a womb, and not only of life, but also of death. The body is not made out of dust (unlike *Genesis* 2: 7), nor does it return to dust (unlike *Genesis* 3: 19), but is made out of land and dies into land. The body is the shell for the spirit. The spirit is the seed within the nutshell from which new life springs.

The land reveals the law as the law for the Rainbow Spirit Elders is 'revealed in the land' (Rainbow Spirit Elders, 2007, p. 11). It also revealed for them in the scriptures of the Bible. The land for the elders is 'like the Scriptures – sacred stories [such as the Rainbow Spirit, the Rainbow Serpent and the Waugal/Waagle] and signs are inscribed on the landscape, and readily available for those who can read them' (p. 20). Secular stories and signs, such as highways, mines, cities and farms, are also inscribed on the land:

> highways have crossed the land and erased the trails of our ancestors. Mines have been dug deep into the land and desecrated our home countries. Cities have been constructed on sacred places where our stories once gave us strength and meaning. Farms have torn up the land where the Creator Spirit once provided plants and animals for our livelihood.
>
> (Rainbow Spirit Elders, 2007, p. 44; see illustration on p. 43)

Highways, mines, cities and farms (not to mention railway lines, power-lines, runways, roads, drains, fences, etc.) can be read by all as colonizing inscriptions in the landscape, as wounds inflicted on the living body of the earth, and not as markers of 'progress', or 'development', or 'nation-building'.

The inflicting of these wounds on the land produces cries of pain and anguish that can be heard by all who have ears to hear them, but so many have blocked their ears so that they cannot hear these cries or anybody else's. For the Rainbow Spirit Elders 'the Creator Spirit is crying because the deep spiritual bonds with the land and its people have been broken. The land is crying because it is slowly dying without this bond of spiritual life' (Rainbow Spirit Elders, 2007, p. 42). The land is also crying because it is 'in slavery to Western forms of exploitation and control' (p. 53), such as colonization. The land needs liberation in a land liberation theology.

Unlike the stories of scripture inscribed in written words of dead black ink on dead white paper, sacred stories are traced in spoken words and painted depictions of/in the trails of ancestral beings ('song lines') and 'life-forces' across the living land (Rainbow Spirit Elders, 2007, pp. 38 and 39) and from the land crying beneath highways, mines, cities, farms, railway lines, power lines, runways, drains and fences. They are there for all to see and hear. The stories of the Rainbow Serpent and the Rainbow Spirit tell a powerful, single story of the land and water, life and death, of ancient and modern ways, of seeing both ways together at the same time in a binocular and microscopic vision that sees two cultures together and what is above and below together, that does not merely see what is on the dead surface of the earth – landscape, private property – but also sees what is below the surface – land, commons – its depths of geology, history and archaeology and its living flows and processes in biology.

Both the Rainbow Serpent and the Rainbow Spirit are also problematic in ways already discussed. The Rainbow Serpent is problematic for the way in which it imposes a unitary, homogeneous and anthropological ur-myth on culturally diverse and geographically dispersed Aboriginal cultures. The Rainbow Spirit is also problematic for white-washing the negative, pejorative connotation of the biblical serpent out of Aboriginal theology. Achieving a rapprochement between the two figures — serpent and spirit — and between the three -ologies could produce a dialogic anthropo/eco/theo/logy, or even better sacrality, between spirituality and materiality that cares for earth. Rather than the -ologies that privilege the logos or word and the -isms in general that exploit and destroy earth (industrialism, militarism, capitalism, neo-liberalism, etc.), and sanctuarism in particular that only cares for special places while the rest of earth goes to hell in a hand-basket, the -ities in general and sacrality in particular cares for earth and all earthly places.

The Symbiocene

The marsh monsters, swamp serpents and fiery dragons encountered in various guises in this and previous chapters of the present book are not only (as we have seen) instances of the ways in which the sacred becomes the monstrous, and even demonic, but are also figures that embody and express earthly sacrality. They should be regarded as figures of reverence and respect that can instil and nurture earthly sacrality and develop ways of living bio- and psycho-symbiotic livelihoods in bioregions of the living earth in the Symbiocene. They are deities to believe in. They are not gods in the Western sense of beings above and outside of earth, but deities in the Eastern sense of beings below and of earth who embody, or body forth, forces, principles and virtues of living and loving.

The Symbiocene is the paradigm of mutuality posed against the failed paradigm of mastery of the anthropocentric and the hoped-for age embracing and following the Anthropocene. The Anthropocene can be defined as the geological age of the laying and layering up and down of anthropogenic strata in the pollution of land, waters and air, and the heating up of all three in the new, disrupted (and disruptive) arrangement of the four elements and the four seasons (see Giblett, 2013, chapter 18). These anthropogenic strata are what Michel Serres calls 'plaque tectonics' in which 'a great many humans form a "plaque", a formation that *disturbs* [the] functioning [of physical communality] [. . .] *plaques* of physical *scoria* of humans [. . .] are encrusted upon and overlap the globe' (cited by Conley, 1997, p. 65; see also Serres, 1995, p. 16). These plaques of humanity are reshaping the globe, like plate tectonics, and 'the Earth is quaking anew', as Serres (1995, p. 86) puts it, in feral quaking zones, as I put it (see Giblett, 2009, chapter 1).

The Anthropocene came out of the economic politics of mercantile capitalism, industrial capitalism, enclosure of the commons, private property, the commodity market and the feral quaking zone. The Anthropocene came out of its drive for, and the failed paradigm of, mastery. The anthropogenic stratigraphic layers of the Anthropocene gave rise to what Bruno Latour calls 'the New Climate Regime' in which 'there is no longer any question of "mastering" nature' (Latour, 2017, pp. 111–115). Rather, 'nature' is mastering, or monstering, us. The new, anthropogenic climate regime rules – and it is *not* okay.

The Symbiocene can be defined as the geological age of the laying and layering up and down of bodies of land, waters, air, beings and things in biogenic and non-biogenic strata, and the inter-corporeal relationship of loving union between them in the feral and native quaking zones (see Giblett, (2009, chapter 1)) and as portrayed in Shane Pickett's paintings. The Symbiocene comes out of the economic politics of the commons, compassion for all beings, mutual aid, the carnivalesque marketplace (not the capitalist market) and the native quaking zone. The Symbiocene comes out of its love for and paradigm of mutuality.

As ever, Thoreau was there before anyone else when he wrote in the mid-nineteenth century with stunning prophetic insight (foresight) in *Walden*, the old testament of conservation, that:

> The earth is not a mere fragment of dead history, stratum upon stratum like the leaves of a book, to be studied by geologists and antiquaries chiefly, but living poetry like the leaves of a tree, which precede flowers and fruit – not a fossil earth, but a living earth; compared with whose great central life all animal and vegetable life is merely parasitic.
>
> (Thoreau, 1997, p. 289)

The earth is not an object to be studied by geologists alone who preside over the dead strata of the Anthropocene and by antiquaries, such as John Muir, who study the dead strata of the fossil earth like leaves in a book holy or otherwise. The earth is not only layers of capitalist crud laid up and down in the land, waters and air from the burning of fossil fuels and the manufacturing and disposal of commodities in the Anthropocene. The earth is also a symbiotic body with a trunk and palms like leaves of a trees, a living poem in the Symbiocene.

The symbiotic body of the earth comes out of the traditional cultures in which earth is body and/or the body is earth. It supersedes the machine body of modern Western medicine, the battlefield body of illness narratives, the grotesque body of the lower strata, the monstrous body of the slimy depths, the Fascist body of the war machine, the textual body of the surface of inscription, the sporting body imprisoned in the time-machine and the cyborg of the body-machine of the civilian soldier. The Symbiocene embraces these bodies in the body of the earth in which healthy bodies, minds, lands and waters can and do flourish in their bioregional and local home place of the living earth (see Giblett 2006; 2008a; 2009, chapter 8). The stratigraphic Symbiocene cuts across the biosphere, the ecosphere, the public and private spheres, the electro-magnetosphere that makes wireless telecommunication possible and the extra-terrestrial sphere where the communication satellites orbit. Transportation and communication technologies function in these spheres. They are the greater ambit of the earth home (see Giblett 2008b; 2011, chapter 2).

Making a connection to local place, its plants, animals and their seasonal changes, is a necessary response, and antidote, to the globalized world in which many people now live and work and which impacts on our lives in numerous ways. It is vital to think and act locally as well as globally. Connecting to local place can be a reclusive retreat into a smaller, narrower and safer world away from the incursions of the bigger, badder global world. But it is also a way of acknowledging and respecting the interconnectedness of all life from the local to the global and back again. Our lives are lived locally (if not also globally) and are dependent on local air, water and food mainly supplied from within and by our bioregional home habitat. We have aerials and cables, but we also have roots – however shallow or transient they may be. We feed off nutrients in the soil and although we may up roots and change soil occasionally or frequently, we are still putting down them into a soil, drinking local water, breathing the air around us and largely eating local food. That air and soil has a history, a human and a natural history. Knowing its composition enriches our lives and helps to connect us to the other living beings living in the same soil. That sense of mutuality between people and place is vital to conserving a place and the planet.[4]

All living creatures – plants and animals, human and non-human – live in a bioregion, a catchment or watershed, and an air-shed with its unique suite of plants and animals. The bioregion is the place of home, the home place. All creatures are dependent and impact on their bioregion to greater or lesser extent, with longer or shorter term damaging and/or rehabilitating effects. The relationship between creatures and their bioregion (and ultimately the earth and the biosphere) takes place on a continuum from the parasitic to the symbiotic through the inquilinistic (in which one party shares the home of another without significant disadvantage to the home-owner, such as normally the embryo in utero; see Giblett, 2011, chapter 12). This continuum is not only biological and bio-geographical, but also psychological and spiritual (see also Giblett, 2011, chapter 12).

The bioregion is home not only to biological creatures, but also to spiritual, even mythological, creatures who are emanations and expressions of the place and its processes. They are the spirits of the place, the *genii loci*. The 'spirit of a place' is a being, not a feeling; it is a way of being, and a way of living. Marsh monsters and swamp serpents and other monstrous figures, such as fiery dragons and earthy/watery alligators and crocodiles, are such creatures. Rather than regarding them with horror and demonizing them as evil beings, they should be respected and revered as creatures of the sacred earth. Long may they live; long may other creatures live with them. Please help conserve them and their homes of places on, under and above the earth. After all, it is their – and our – only home.

Notes

1 I retain the varying spellings of Waugal/Waagle/Wagarl from the sources I cite partly out of respect for them and partly to stand as testament to a living, non-alphabetic culture. They can all be pronounced as 'woggle'.

2 The Aboriginal Nyoongar people of south-western Australia have rejected Mudrooroo's identity as an Aboriginal person. By quoting from his work I do not accept (or deny) his aboriginality. It is not up to me to decide. I do not cite his work in this book as that of an Aboriginal writer or person, or not. I quote his work as it is invaluable and as his contribution to cross-cultural dialogue and understanding is enormous.

3 Paul Kingsnorth (2017, p. 272) has referred critically to 'our manifest destiny as a chosen species'.

4 For my bioregional connection to the local place where I lived for thirty years, see Giblett (2006; 2013).

References

Conley, V. 1997. *Ecopolitics: The Environment in Poststructuralist Thought*. London and New York: Routledge.

Eliade, M. 1958. *Patterns in Comparative Religion*. R. Sheed, trans. London: Sheed and Ward.

Flood, J. 1995. *Archaeology of the Dreamtime: The Story of Prehistoric Australia and its People*. Revised edition. Sydney: HarperCollins.

Giblett, R. 2006. *Forrestdale: People and Place*. Bassendean: Access Press.

Giblett, R. 2007. Black and white water. *In*: E. Potter, A. Mackinnon, S. McKenzie and J. McKay, eds, *Fresh water: New perspectives on water in Australia*, Carlton: Melbourne University Press, pp. 31–43.

Giblett, R. 2008a. *The Body of Nature and Culture*. Basingstoke: Palgrave Macmillan.

Giblett, R. 2008b. *Sublime Communication Technologies*. Basingstoke: Palgrave Macmillan.

Giblett, R. 2009. *Landscapes of Culture and Nature*. Basingstoke: Palgrave Macmillan.

Giblett, R. 2011. *People and Places of Nature and Culture*. Bristol: Intellect Books.

Giblett, R. 2013. *Black Swan Lake: Life of a Wetland*. Bristol: Intellect Books.

Giblett, R. and Webb, H. 1996. Living Waters or Useless Swamps? *In*: R. Giblett and H. Webb, eds, *Western Australian Wetlands: The Kimberley and South-West*, Perth, Western Australia: Black Swan Press/Wetlands Conservation Society, pp. 1–8.

Kingsnorth, P. 2017. *Confessions of a Recovering Environmentalist*. London: Faber and Faber.

Latour, B. 2017. *Facing Gaia: Eight Lectures on the New Climate Regime*. Cambridge: Polity.

Little, L. 2004. *The Mark of the Wagarl*. Illustrated by J. Lyndon. Broome: Magabala.

Maddock, K. 1978. Introduction. *In*: I. Buchler and K. Maddock, eds, *The Rainbow Serpent: A Chromatic Piece*, The Hague: Mouton, pp. 1–21.

Mountford, C. 1978. The Rainbow-Serpent Myths of Australia. *In*: I. Buchler and K. Maddock, eds, *The Rainbow Serpent: A Chromatic Piece*, The Hague: Mouton, pp. 23–97.

Mudrooroo 1996. A Snake Story of the Nyoongah People: A Children's Tale. *In*: R. Giblett and H. Webb, eds, *Western Australian Wetlands: The Kimberley and South-West*, Perth: Black Swan Press/Wetlands Conservation Society, pp. 33, 36–37.

Noonuccal, O. and Noonuccal, K. 1988. *The Rainbow Serpent*. Canberra: Australian Government Publishing Service.

O'Connor, R., Quartermaine, G. and Bodney, C. 1989. *Report of an Investigation into Aboriginal Significance of Wetlands and Rivers in the Perth-Bunbury Region*. Leederville: Western Australian Water Resources Council.

Pickett, S. 2003. Waagle – Rainbow Serpent 1983. *In*: B. Croft with J. Gooding, *South West Central: Indigenous Art from South Western Australia 1833–2002*, Perth: Art Gallery of Western Australia, p. 54.

Pickett, S. 2004. Waagyl and Yondock Story. In: *Walyalup Dreamings*, Fremantle: City of Fremantle, p. 9.

Radcliffe-Brown, A. R. 1926. The Rainbow-Serpent Myth of Australia. *Journal of the Royal Anthropological Institute of Great Britain and Ireland*, 56, pp. 19–25.

Radcliffe-Brown, A. R. 1930. The Rainbow-Serpent Myth in South-East Australia. *Oceania*, 1, pp. 342–347.

Rainbow Spirit Elders, The. 2000. *The Rainbow Spirit in Creation: A Reading of Genesis 1*. N. Habel, ed. and trans. Collegesville, Minnesota: Liturgical Press.

Rainbow Spirit Elders, The. 2007. *Rainbow Spirit Theology: Towards an Australian Aboriginal Theology*. Revised or second edition. Adelaide: Australian Theological Forum Press.

Serres, M. 1995. *The Natural Contract*. E. MacArthur and W. Paulson, trans. Ann Arbor: University of Michigan Press.

Thoreau, H. 1997. *Walden*. Boston: Beacon. First published 1854.

Webb, H. 1996. Aboriginal Country: Not a Construction, A Way of Being. *In*: R. Giblett and H. Webb, eds, *Western Australian Wetlands: The Kimberley and South-West*, Perth, Western Australia: Black Swan Press/Wetlands Conservation Society, pp. 61–76.

Index

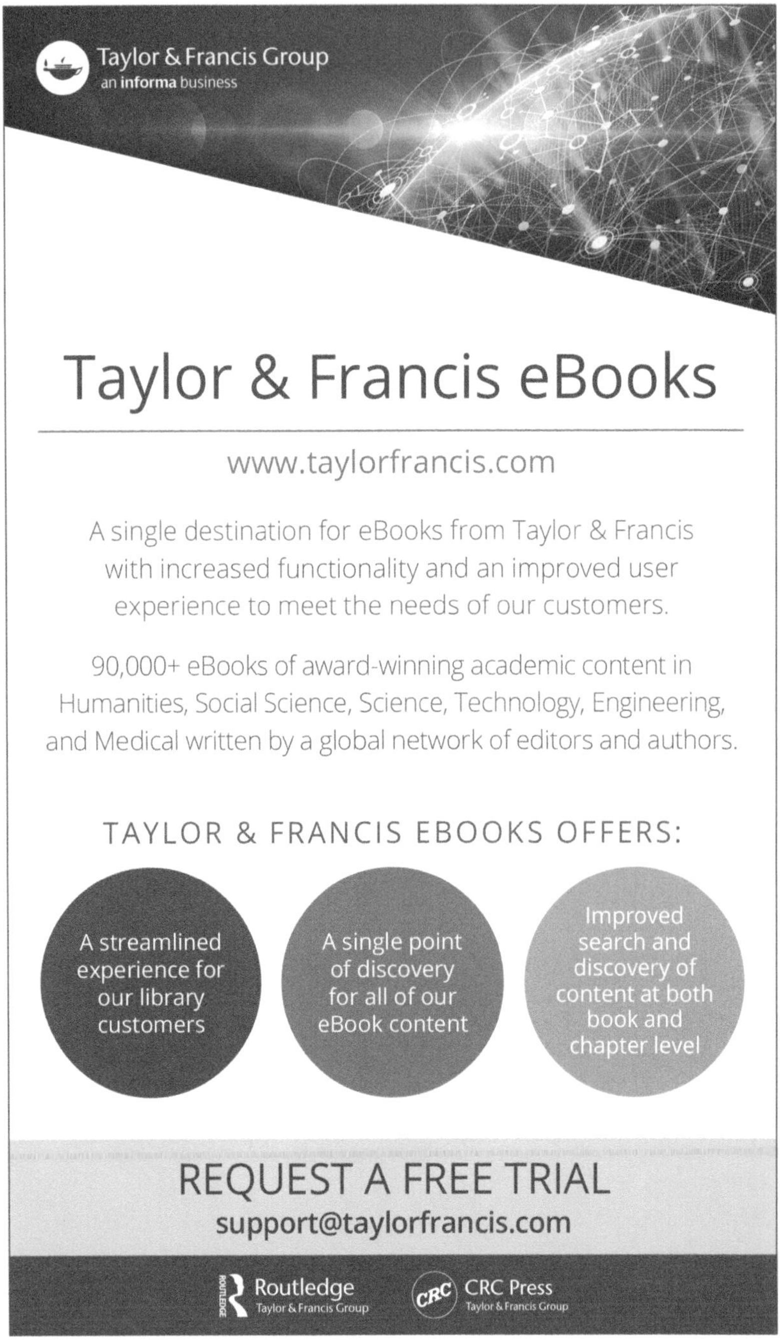
Taylor & Francis Group
an informa business
Taylor & Francis eBooks
www.taylorfrancis.com
A single destination for eBooks from Taylor & Francis with increased functionality and an improved user experience to meet the needs of our customers.
90,000+ eBooks of award-winning academic content in Humanities, Social Science, Science, Technology, Engineering, and Medical written by a global network of editors and authors.
TAYLOR & FRANCIS EBOOKS OFFERS:
A streamlined experience for our library customers
A single point of discovery for all of our eBook content
Improved search and discovery of content at both book and chapter level
REQUEST A FREE TRIAL
support@taylorfrancis.com
Routledge
Taylor & Francis Group
CRC Press
Taylor & Francis Group

For Product Safety Concerns and Information please contact our EU representative GPSR@taylorandfrancis.com Taylor & Francis Verlag GmbH, Kaufingerstraße 24, 80331 München, Germany